A GARDENER'S GUIDE TO PROPAGATING FOOD PLANTS

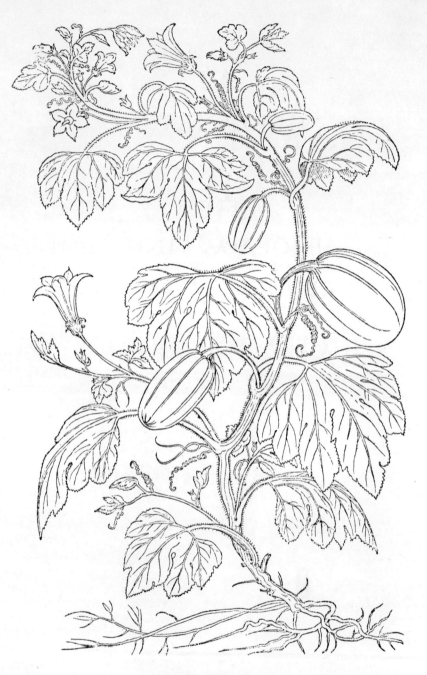

Pumpkin-*Cucurbita pepo*

A GARDENER'S GUIDE TO PROPAGATING FOOD PLANTS

FRANKLIN HERM FITZ

CHARLES SCRIBNER'S SONS
New York

*This book is
dedicated
to all those who seek
a fuller measure of
self-sufficiency*

Copyright © 1978, 1983 Franklin Fitz

Library of Congress Cataloging in Publication Data
Fitz, Franklin Herm.
  A gardener's guide to propagating food plants.

  Bibliography: p.
  Includes index.
   1. Plant propagation.   2. Food crops.   3. Gardening.
I. Title.
SB119.F57   1983   631.5'2   82-21409
ISBN 0-684-17655-6

An earlier version of this book was published in 1978 under the title *Compleatly
Self-Sufficient Food-Plant Propagation.*

This book published simultaneously in the
United States of America and in Canada—
Copyright under the Berne Convention.

All rights reserved. No part of this book
may be reproduced in any form without the
permission of Charles Scribner's Sons.

1 3 5 7 9 11 13 15 17 19 F/C 20 18 16 14 12 10 8 6 4 2

Printed in the United States of America.

Contents

PART ONE
Methods of Plant Propagation
vii

A Brief History of Plant Propagation	1
1 *Sexual Methods: Seed Production*	4
Annual, Biennial, or Perennial?	8
Gathering Seed	10
Testing for Seed Viability	11
Stratification	12
2 *Asexual Methods: Plant Division*	13
Cuttings	14
Crown Division	16
Root Division	17
Suckers	17
Stolons (Runners)	17
Layering	19
Tip Rooting	21
Bulbs	21
Tubers	22
Bulblets	23
3 *Asexual Methods: Budding and Grafting*	24
General Principles	25
Growing Your Own Rootstock	26
Budding	27
Grafting	30
4 *Selecting the Parent Plant and Other Tips on Plant Propagation*	36

PART TWO
Propagating Specific Food Plants
39

Plant Hardiness Zone Map	41
Botanical Classification of Common Food Plants	137
Bibliography	145
Index	149

PART ONE

METHODS OF PLANT PROPAGATION

A Brief History of Plant Propagation

From his ancestors, modern man has inherited a rich and diverse legacy of food plants. The cultivation, selection, and propagation of fruits, nuts, leafy greens, roots, grains, succulent shoots, tubers, bulbs, and seeds is our heritage, something we now take for granted. But most of us never wonder about the secrets of plant reproduction. At a time when some degree of self-sufficiency may very well be a key to the future, we might consider propagating our own food plants as one step toward that goal.

Two hundred years ago, most gardeners were still propagating their own plants, either by seed or by vegetative means. Their practices had begun in the earliest days of plant domestication and by that time were considered routine. But they began to die out with the coming of our age of specialization. As early as the birth date of the United States, seed growing had become an industry. In 1795, George Washington was prompted to write from his home in Mount Vernon, "It is shameful for gardeners and farmers to be buying seeds that their own soils and climates will produce."

The trend toward buying plant propagules—seeds, tubers, bulbs, and so on—increased through the years, especially as plant breeding succeeded in improving such crop characteristics as yield, hardiness, and resistance to drought and disease. By the early 1900s, gardeners were actually discouraged from saving seeds. Bailey (1910) stated that growing seeds "requires expert knowledge of soils and climate and methods of handling every kind of crop." For economic reasons, too, he claimed, "The cost of seed is ordinarily a trifling matter in comparison with the expense of the season's labor and the value of the crop."

This kind of thinking was prevalent until the near-present, when Thompson and Kelly (1957) dealt the death blow to seed savers: "Seed growing is a highly specialized business requiring particular knowledge and skill not possessed by many vegetable growers: therefore most gardeners should buy their seeds through regular channels."

Once again, as with tanning hides and making cheese, we were told not to bother trying since the experts could do it better. Kains (1935) must have agreed; in a book devoted entirely to independence on the land, in regard to saving seeds he stated, "There are too many risks to run—leave it to the specialists." Steelman (1951), in a similar book, completely neglected the subject. Stout (1971) discouraged any attempt to save seed at all.

Fortunately, the lore of plant propagation did not die, to be relegated to a chosen few. In what seems to be the beginnings of a countermovement, Nissley (1942) has encouraged seed savers, devoting an entire chapter to the methodology. More recently, many authors have resounded the call, and their cries are entirely justified.[1] A few simple generalizations about plant life cycles and

1. See Meeker 1969; Larger 1972; Shade 1974; Anonymous 1975; Emery 1975; Moon 1975; Moore and Moore 1975; Douglass 1976; Johnston 1976, 1977, 1981; King 1977; Mariner 1977; Miller 1977; Willmann 1977; Anonymous 1978a; Rogers 1978.

reproduction can teach even the novice how to save seed successfully from his garden plants as well as how to propagate them by vegetative methods.

It is true that saving seed can be exacting work, competing for critical time during harvest periods. A gardener must also have some knowledge of plant breeding, of the techniques for growing seed crops, and of curing and handling seeds. He must be able to observe and judge, comparing types of plants and their performances. Gardeners must be aware, too, of hybrid "dropouts" and possible crosses that would result in undesirable offspring. But these factors are easily overcome by the interested gardener, and the benefits to be gained are immense:

1. Selective home plant propagation, including seed saving, can improve the yield and quality of the crop. Those seeds saved will fit your particular soil type and local climate, becoming better adapted with each generation.
2. You can save money. At today's prices, the cost of seed and other propagules is no longer insignificant.
3. For an out-of-the-ordinary variety that may be difficult to find, seed saving may be your only alternative.
4. You will know your plants' type, origin, and parentage.
5. Your home-propagated plants and seeds will, naturally, be organically grown and untreated.

Home plant propagation is not only an enjoyable challenge but is also a way of getting closer to a "grass-roots existence" and of working hand in hand with mother nature.

1

Sexual Methods: Seed Production

We tend to think of reproduction in terms of what is most familiar—in the higher vertebrate animals, in the fertilization of a female egg by a male sperm to produce the cell that will develop into the new animal. Plants propagate this way too, essentially, and the seed is the result. But there are many nonsexual methods of reproduction often employed by plants and the lower invertebrate animals as well. We might sort these methods into two general classes: sexual, involving a male and a female; and asexual, involving a single parent.

The seed of a flowering plant is a package containing the dormant embryonic plant of the next generation and a stored food supply, all wrapped in a protective seed coat. Flowers are reproductive organs, each usually containing protective sepals, colorful petals that attract pollinators, stamens, and a central pistil. Stamens

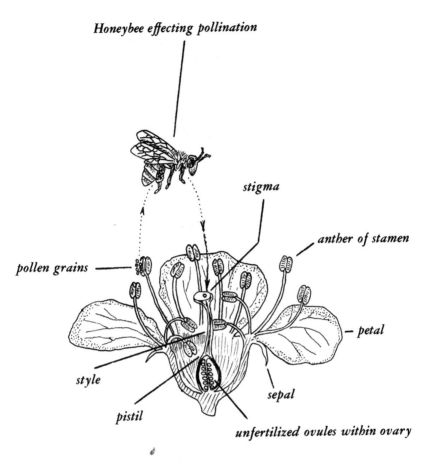

Figure 1. A generalized flower, showing the essential sexual parts.

are the male part of a flower and produce pollen grains, which act like sperm. The pistil is the female part and contains ovules in the protective ovary. Ovules are somewhat like eggs and are eventually fertilized by the sperm nucleus of the pollen grain. In all cases, pollen reaches the sticky tip of the pistil (the stigma) and grows down to the ovary, where it fertilizes the ovule. The wind, insects of many kinds, or birds generally accomplish this pollination (Fig. 1).

The fertilized ovule develops into the seed, and the ovary wall of the pistil usually becomes the fruit. Most flowering plants bear flowers containing both stamens and pistils and are called *perfect*. It is sometimes possible for the pollen from the stamen of one flower to fertilize the ovules of the pistil within the same flower (as in tomatoes, peas, and beans). This is called *self-pollination*. Often no external pollinating agent is required. But this is a bit like inbreeding, and more often the pollen must reach another flower, either on the same plant or on a different plant of the same species, to fertilize effectively. Most plants use this latter method, called *cross-pollination*. Some cross-pollinators, such as cabbage, require that the pollen from one flower reach the stigma of a flower on a different plant of the same species; such plants are termed *self-incompatible* or *self-sterile*. In this case, viable seed may be produced only when two or more plants are involved.

To complicate matters further, plants may divide into sexes, bearing female flowers—having pistils but no stamens—and male flowers—having stamens but no pistils. If flowers of both sexes are borne on the same plant, as in corn, persimmon, and all members of the gourd, walnut, and beech families, the plant is called *monoecious*. But when plants bear female and male flowers separately, as on spinach, yam, papaya, and asparagus, they are then called *dioecious*.

To grow seed one should know whether a plant is self- or cross-pollinated, self-compatible or incompatible, perfect, monoecious, or dioecious, and the pollinating agent required, wind or insect. This information helps in determining the numbers and types of plants to set out for seed and the spacing required. Insect-pollinated plants must be separated more widely than wind-pollinated plants to prevent cross-pollination between varieties.

Sexual propagation necessarily passes genes from each parent to the seed and may or may not result in offspring similar to the parents. *True-breeding* plants, often self-pollinated, are inbred sufficiently so that each parent contains genes very similar to the other's, keeping the offspring true to type. Other plants, while appearing similar, may contain variable genes and will produce offspring that differ among themselves and from their parents. Although this phenomenon of variability is essential to keeping mother nature's populations fit to their environments, it is not desirable in selected crop plants. Crossing between varieties of a species should be prevented to avoid untrue hybrid offspring. In some cases, crossing between different species within the same genus may occur; these potential crosses should also be prevented. Intergeneric crosses seldom occur, except among grains and usually by forcing. As a general rule, one need never worry about cross-pollination taking place between two plants in different genera (such as beets and spinach). Because it is important to know what possible crosses do exist, some understanding of the classification of garden food plants will be helpful when planting a garden for seed production. The Botanical Classification of Common Food Plants (page 137) lists the plants presented in Part Two of this book according to their botanical families, genera, species, and varieties. To see how this table can work, consider the many old wives' tales about cucurbits intercrossing, many of which are myths (Schales 1969, Hawthorn 1961, Deakin, Bohn, and Whitaker 1971). The Botanical Classification reveals that cucumbers are in a different genus from watermelon and do not cross with it. Figure 2 gives other examples within this group, showing possible crosses and negating others. Using the Botanical Classification, similar analyses can be made for the crucifers (mustard family) or any pair of plants.

You may find that you wish to propagate plants that are F-1 hybrids—the offspring of two distinct, carefully inbred lines containing different genes for most of their traits. When these plants form gametes (by the process of *meiosis*) during reproduction, they essentially "shuffle the genes and deal" or "roll the genetic dice," and the offspring may either revert to one or the other parental type or become a freakish combination of both. A tiny percentage, purely by chance, may resemble the original F-1 hybrid, but we're interested in greater numbers of true offspring. So F-1 hybrid seeds can be considered "dropouts" from the propagation school, and one should not attempt to grow their seed. Many varieties of corn, squash, and cucumber are specially bred hybrids and can be produced only by crossing carefully bred parental stocks. This practice may be beyond the desires of the home gardener, and in this special case the only way to continue raising such plants is to buy the seed.

ANNUAL, BIENNIAL, OR PERENNIAL?

Now let's get down to business! How do we go about growing our own seed? First we need to know one more thing—whether the plant is an annual, a biennial, or a perennial.

Annual Plants

Annual plants live for one year only, growing from seed in spring, flowering later in summer, and producing seed for the next generation that same season. Following seed production, which is usually over by late fall, the plant dies. Lettuce, beans, and peas are good examples.

Biennial Plants

Biennial plants live for two years. During the first season they grow only vegetatively, producing roots, stems, and leaves but no flowers, fruits, or seeds. Examples are carrots, beets, Swiss chard, and many of the crucifers. These plants may be overwintered, either in the garden or in the root cellar, and will flower, fruit, and produce seed during the second growing season, after which they die.

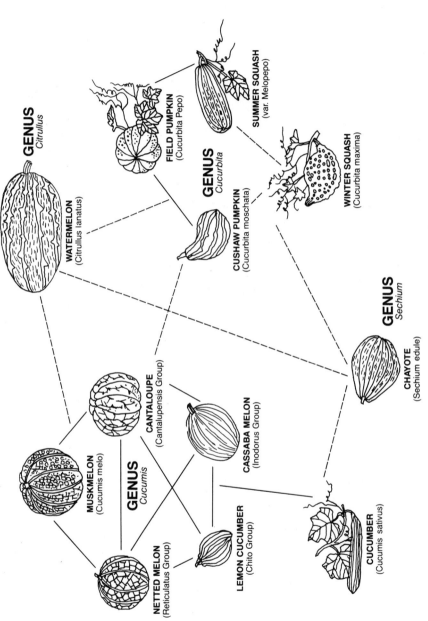

Figure 2. Possible crosses within the gourd family (Cucurbitaceae). Crosses are indicated by solid lines; dotted lines indicate crosses that are not usually possible.

Perennial Plants

Perennial plants live for more than two years. Depending on the plant, vegetative growth may last from one to several years before flowering ensues. At the onset of flowering, and of the resultant seed production, plants may produce fruit and seed for a few to many years before expiring. Perennials often lend themselves to propagation by methods other than, or in addition to, seed production.

GATHERING SEED

To collect seed the gardener needs simply to foster the growth, flowering, and seed production of each plant he desires to propagate, ensuring proper pollination and no undesirable crossing. Then, at the time of maturity and ripeness of the fruit and seed, he must collect the seed according to the method appropriate to each plant. (Part Two provides specific details on each plant.) Some fruits are dry and yield fairly dry seeds that are easily separated by shaking, threshing, screening, and winnowing (all of the members of the mustard, sunflower, goosefoot, grass, pea, carrot, and buckwheat families, for example). Other fruits are fleshy and moist (as in some members of the tomato, lily, gourd, and other small families) and often require separation by water or even a short period of fermentation to remove sticky, mucilaginous materials from the seed. Whatever the case, you'll probably find some materials handy in your seed-saving exploits:

1. Window screens in wooden frames for drying seeds
2. Hardware cloth ($\frac{1}{8}$-inch and $\frac{1}{4}$-inch mesh) in frames for sieving seed
3. Smaller sieves, if possible, for sieving tiny seeds
4. Jars and envelopes for storage
5. Grocery sacks and canvas for drying seeds
6. Cookie sheets for winnowing

7. Crock, glass jars, or wooden vessels for fermentation when required.

Reject puny, shriveled seeds; they may not be fully developed and will germinate poorly. Once you've collected and thoroughly dried your seeds, carefully label each package as to type and date of collection and store them in a cool, dry place (Carl 1975) protected from insects, moisture, and heat. Most seeds will keep for at least one year and others will keep for several years. (Part Two lists specific viabilities.) Different authors present different seed viability times; those presented here are "best estimates" taken from several sources.

TESTING FOR SEED VIABILITY

Before planting seed each spring, it is important to know the viability, or rate of germination, you may expect. This will affect the density at which you sow the seed.

For most seeds, testing for viability is quite simple. It is important to provide the seeds with moisture, warmth, and oxygen in such a way that fungal growth is discouraged. Place a counted number of seeds on a clean, damp cloth (not soaking wet—it needs to "breathe"), roll it up, and place it in a warm place, such as a bowl by the stove. This cloth should be moistened and inspected daily. After some time, the number of seeds that sprout can be counted and the percentage calculated.

Some seeds sprout within a day or two, especially beans, peas, corn, and radishes; most common garden vegetables germinate within a week. Others, notably members of the carrot family (parsley, dill, carrots) and the annual peppers (*Capsicum annuum* groups) require a couple of weeks to a month or more. These slow-to-germinate seeds may best be tested in shallow flats of loose, sandy potting soil to reduce the danger of decay. Plant a known number in a marked row, keep them watered, and count how many eventually appear.

STRATIFICATION

Another problem in germination, especially for temperate-zone fruit and nut trees and berry-producing shrubs, is seed dormancy. To break dormancy, the seeds must be "stratified" in pans of moist sand or peat at near-freezing temperature (33°F to 44°F), conditions that simulate overwintering in nature. Place a layer of damp sand about one inch deep in a flat container with small drainage holes in the bottom. Lay the seed on the sand (small seeds should be wrapped in a layer of cloth), and spread another layer of sand over the seed. Alternate layers of sand (or peat) and seed until the flat is filled. The stratified seeds must be kept cool and moist until planting time. The box may be buried outdoors in well-drained soil to a depth of six to eight inches, or it may be placed in a cool cellar, if normal winter temperatures remain sufficiently cold for a month or more. In warmer climates the box may be placed in the refrigerator. Seeds stratified outdoors should be adequately protected from possible rodent damage by screens of ¼-inch-mesh hardware cloth. Specific cold periods and temperatures for common fruit and nut rootstock seeds are given on page 26, "Growing Your Own Rootstock." Following stratification, the seeds may be tested for viability as described above.

Once the seed viability is known, planting density may be determined. For example, for a seed batch with 100 percent viability, plant the number of seeds equal to the number of plants desired (perhaps allow a few extra). If viability is 50 percent, plant seeds about twice as thick as the number of desired crop plants. And, of course, if viability is at or near 0 percent, do not plant the seeds at all!

Asexual Methods: Plant Division

Asexual methods of plant propagation, involving only one parent, are entirely different from the sexual methods: a gardener always (except in rare mutant forms) gets the same plant he starts with, a plant genetically true to type—literally a "chip off the old block." The reason for this fidelity is simple—the plant is reproducing a part of itself with nothing more than a simple multiplication of its own cells (the process of *mitosis*), which produces more cells with chromosomes and genes identical with those of the parent cells. There is no probability or chance involved. All asexual methods, from runners and grafting to bulbs and tubers, can be counted upon to reproduce faithfully the desired variety; in fact, new varieties cannot be formed efficiently by these methods. Asexual methods of propagation are discussed here in general; the specific methods applicable to each food plant are indicated in Part Two.

CUTTINGS

Cuttings are portions clipped from the stem or root of the parent and transplanted to pots or rooting beds where they develop all the parts of a new and whole plant.

Green Cuttings

Stems of many nonwoody plants may be clipped from the parent, along with one or more leaves, and inserted into loose, rich, sandy soil, where they will develop roots over a period of several weeks to a few months (Fig. 3). Take care to prevent the cutting from drying out. To reduce water loss, clip the leaves in half, especially if they are large. This method is especially applicable to ornamental, flowering garden plants, but can be used successfully on tomatoes and other herbaceous plants. Green cuttings are made while the plant is actively growing.

Hardwood (Dormant) Stem Cuttings

For woody vines and other tree or shrub plants (grape, banana, currant, date, fig, filbert, gooseberry, mulberry, olive, quince, pineapple, and pomegranate), cut the younger but mature stems

***Figure 3.** A green cutting from a tomato.*

into sections, the base of the cutting being made just below a node where buds appear and the top slightly above a bud. Make cuttings in fall, winter, or early spring, any time between the onset of fall dormancy and spring bud-break. Cuttings may vary from 6 to 12 inches long or longer. Tie into bundles (all tops at one end, usually cut flat; all bottoms at the other end, usually cut slanted) and store in the root cellar or buried outdoors in a horizontal position in moist soil, sand, sphagnum peat, or a mixture thereof. Store during the first half of the winter (from two to four months) at 50°F to 55°F to induce callus growth, which facilitates later root development. Store cuttings for the remainder of the winter at 35° to 40°F. This second, cooler period retards growth until spring and also serves to break the dormancy in the buds. Plant these cuttings in the spring as soon as soil and weather conditions are favorable. Roots will then grow from the base of the stem, and the buds will expand to form the new leaves.

Cuttings may be greenhouse-grown during the first season or started directly in a permanent location.

Figure 4. A hardwood cutting (dormant) of gooseberry.

Figure 5. A root cutting of horseradish with new leaf growth.

Root Cuttings

For some plants, especially the blackberry and horseradish, cut pieces of root three to four inches long, and overwinter as you would hardwood cuttings. Select roots of pencil thickness, and cut after growth ceases in autumn. Plant these roots in the spring, and each will develop into a new plant (Fig. 5).

CROWN DIVISION

The crown of a plant is the portion where the root joins the stem. Cut the crown longitudinally, allowing a portion of stem and root on each piece, and plant it upright in the ground in the fall, or overwinter it in the root cellar until spring before planting. New stems will sprout from the top of the crown and new roots from the bottom, thus producing several plants from a single parent. This method is successfully employed with rhubarb.

ROOT DIVISION

In the fall, when the root system (of asparagus, for example) contains next year's buds, which develop by fall but remain dormant until spring, dig up the roots as soon as the plant dies back. Cut the root system into chunks, each containing a mass of roots with dormant perennating buds (buds on the underground rootstock, which will develop into stems with leaves, flowers, and fruits the following year). Transplant each piece immediately to its new location, and in the spring several new plants will appear. It may take a year or so for the new plants to become fully robust and bear fruit. Root division is not only a means of propagation but also a thinning mechanism that is useful when plants become congested.

SUCKERS

Shoots develop from the roots or underground stems of a parent plant and become young plants around it. Globe artichokes (Fig. 6a), banana, blackberry, and raspberry (Fig. 6b) are good examples. Simply cut the sucker, together with its developed roots, from the parent plant, and transplant it under conditions of high humidity. This method relieves congestion while providing a source of new plants.

STOLONS (RUNNERS)

Trailing and reclining shoots that root periodically are called stolons. The strawberry, which is propagated almost exclusively by this method, is the best example. Where the tip of the stolon roots, a new leafy stem also appears (Fig. 7) and, when it is well rooted, it can be cut from the parent branch and transplanted to a new location. The new plant is already complete, with a stem and root

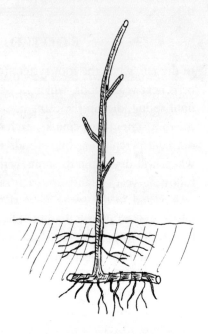

Figure 6a. A sucker from a globe artichoke detached from the parent.

Figure 6b. A sucker from a red raspberry plant, still attached to parent plant but growing new roots.

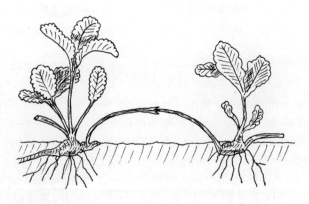

Figure 7. Strawberry plant reproducing by stolons.

system, and needs only to grow further to produce flowers and fruit.

Left to their own, these new plants will congest the original bed and must be removed. Their removal will provide a source of new plants and also allow the parent plants space to grow.

LAYERING

One of the easiest and most successful methods of plant propagation is layering, or rooting stems in soil while they are still attached to the parent plant. Success is due to the fact that the stem is still receiving water and nutrition from the parent while it develops its own root system. Often a full year is required before sufficient root systems are formed, thus allowing the new plants to be cut from the parent. Layering is practiced most often from late spring to midsummer.

Simple Layering

This method is applicable to woody shrubs with flexible stems. Simple layering is similar to tip-rooting and may be used on any woody shrub with branches flexible enough to be bent to the ground. Black raspberry is often simple layered and blueberries and currants may be too.

Bend the stem until the portion just behind the tip rests on the soil. The part of the cane in contact with the soil may take root with no help, but covering the stem behind the tip with soil will encourage it. Root development, especially in hardwood plants, can be further facilitated by notching the underside of the stem where it is in contact with the soil. If the stem springs up or is difficult to hold down, use rocks, or even pegs, to maintain soil contact (Fig. 8). Generally allow a full season of growth before separating the new plant from its parent.

Figure 8. Simple layering.

Mound Layering

For plants with stiff stems that would break if arched to the ground (blueberry, currant, gooseberry), use this method. First wound each stem base slightly with a knife or shovel to induce better rooting. Then mound soil up around the plant and between the stems, covering the wounds. Roots will develop from the covered, wounded portion of the stem (Fig. 9). After one or two seasons' growth, when roots have developed sufficiently, separate the new plants, and transplant each to a new location.

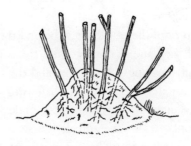

Figure 9. Mound layering.

Compound (Serpentine) Layering

A modification of simple layering, whereby a long trailing stem (such as a grapevine) is covered with soil at several nodes, can be used to develop several new plants from a single stem. Treat the plant exactly as in simple layering, mounding at multiple spots along the stem. Be sure to leave a bud between each soil mound.

TIP ROOTING

Cane plants with arching stems, such as the blackberry, may touch their stem tips to the ground, where they will take root (Fig. 10). Simply cut the newly rooted plants from the parent and transplant them to their new sites.

BULBS

Bulbs are swollen underground portions of the plant stem. They contain root "buds" which, under favorable temperature and mois-

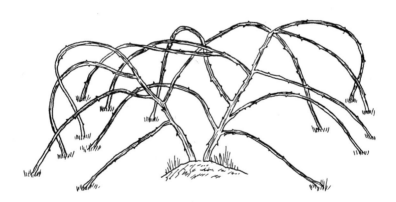

Figure 10. Tip rooting.

Figure 11. An onion bulb producing a young bulb at its base.

ture conditions, produce new roots that allow stem growth. Many bulbs produce smaller bulbs at the base (Fig. 11). These can be harvested in the fall, cured in the sun, and stored in a cool, dry place. Some can be eaten, but a few may be planted in the spring to produce the next year's crop. Shallots, multiplier onions, and some garlics are good examples.

TUBERS

A fleshy underground stem, such as the potato, is called a tuber; it bears "eyes," or stem and root buds, which will elongate in the spring to produce new plants. The harvest should include enough tubers for both eating and planting the following year's crop. In the spring, cut the tubers into pieces (or leave them whole), each with one or more "eyes" (Fig. 12). Plant each piece as you did the

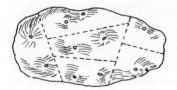

Figure 12. A potato tuber divided into pieces for propagation.

year before. You may dry the cut tubers in the sun for a few hours to form a crust over the fresh cuts before planting in order to aid in the inhibition of rot. Potatoes and Jerusalem artichokes may be propagated by this method year after year.

BULBLETS

Some plants produce a cluster of bulblets at the top of the stalk in place of the usual umbel of flowers and seeds (Fig. 13). In such cases, simply harvest the bulblets in the fall, cure them in the sun for a few days, and store under cool, dry conditions that winter. The following spring, use these bulblets as propagules. Top or tree onions and some garlics propagate by this method.

Other asexual methods of plant reproduction do exist, but they are either experimental (growing a complete carrot from a single root cell) or of no application to food plants (plantlets in leaf notches, corms, air layering).

Figure 13. A cluster of bulblets on a top onion.

Asexual Methods: Budding and Grafting

More than any other methods of plant propagation, budding and grafting are approached with doubt and uncertainty. However, in the words of Talbert (1946):

> The art of propagating fruit trees by grafting and budding has been practiced for more than 2,000 years. In ancient times the attempt was made to shroud it in mystery. The average fruit grower was led to believe that one had to be endowed with special faculties to be able to propagate fruits successfully. In fact, a touch of magic appeared necessary to grow buds of the pear upon the apple, or buds of the peach upon the plum. But now graftage of fruit trees is widely understood, and the methods are so simple that anyone of average intelligence can perform them successfully.

Budding and grafting are applied most often to woody trees, especially the nuts and the pome, stone, and citrus fruits. Both of these methods of propagation involve the union of one part of a plant with another and are practiced for several reasons:

1. To perpetuate various fruit or nut varieties that have appeared by mutation since these plants may only rarely come true from seed
2. To convert an undesirable or unfruitful seedling tree to a suitable type
3. To repair damage by wind, rodents, goats, or farm equipment
4. To grow a number of varieties of fruit tree in a small area by intergrafting two or more fruit types to a single tree.

GENERAL PRINCIPLES

Budding and grafting are practically the same thing, except that in grafting a stem is joined to another plant and in budding a bud is joined. It is important to realize that, if either of these methods is to be successful, the living, growing tissues (xylem, phloem, and cambium) of the joined parts must be fit closely together and kept from drying out. Found just beneath the bark of trees, this triple layer is the "lifeline" that provides the tree with a continuous flow of water and nutrients and allows continued lateral growth. The xylem, nearest the central wood, conducts water upward; the cambium, in the center, allows growth by dividing into new cells; and the phloem, on the outside just beneath the bark, conducts the sugars manufactured in the leaves (along with other substances) to the roots and the rest of the plant.

Budding and grafting are generally limited to woody plants having this living layer beneath the bark. If these "lifelines" of the grafted or budded parts are not brought into smooth contact, growth will be impossible. The process is most successful for plants of a close botanical relationship (see the Botanical Classification of Common Food Plants), either all in the same species (such as varieties of apple), or in the same genus (lemon and lime), or at least closely related within the same family (apple, pear, and quince). The stone fruits, of the genus *Prunus*, are generally compatible, and the plum is easily intergrafted with the peach.

Grafting is performed in order to produce a plant with a superior-bearing top on a superior (vigorous, disease-resistant) root system. The seedling stock on which the graft or bud is placed is referred to as the *rootstock*. The stem, taken from some desired variety of superior-bearing tree, is referred to as the *scion*. Buds are also taken from superior-bearing trees. In budding and grafting, a few simple tools and supplies are necessary:

1. A very strong, sharp, clean knife
2. String, yarn, or rubber bands
3. Chisel or grafting tool (see Fig. 17a) for cleft grafts
4. Grafting wax (or make your own from four parts resin, two parts beeswax, and one part tallow), heated, mixed thoroughly, and applied when just above the melting point.

GROWING YOUR OWN ROOTSTOCK

You may grow your own rootstock from seeds collected from the wild fruits of the type you desire. In most cases, you can simply collect the fruits of the wild seedlings or crab apples. Wash them and allow them to dry for a few days in the open air. Since most fruit seeds will not germinate until they have been chilled at 33° to 46°F in moist conditions for a month or more, you may need to stratify them (see page 12 for instructions on stratification) by burying them in cold sand. Citrus stock seed, however, may be planted immediately, and cherry seeds should not be dried but stratified at 40°F for 110 to 120 days. Air-dry almond seeds and store until spring, then plant for seedling stock. Stratify pecans and peach pits for 100 to 120 days at 40°F. Often peach and cherry seeds can be removed from the fruit and planted immediately, the embryo skipping any dormancy period. Apple seed also may either be cleaned and stratified or planted immediately. Rootstocks are generally grown for one or two years before budding or grafting upon them. Since growing rootstock is a somewhat variable process, I suggest you read Cochran, Cooper, and Blodgett (1961) before starting. Some rootstocks are preferable to others, and although selecting the appropriate rootstock does not fall within the scope of this book, the subject is discussed in great detail by Bailey and Bailey (1976), Talbert (1946), and USDA Forest Service (1974).

BUDDING

Buds for the next season's growth are mature by mid to late summer (July through September), and budding may be performed after that time, when the bark peels fairly easily. Mature buds are dormant, requiring a period of cold before breaking, and will not elongate until the following spring. In all cases, the bud must be grafted in place, cambium layers in close contact and tied securely. All wounds and cuts should be sealed with grafting wax. You can refer to the sketches of the following self-explanatory bud-graft methods (Figs. 14a–b). Bud grafts should be placed as near to the ground as possible to reduce the number of undesirable rootstock suckers that will arise from below the graft. Budding is applicable to all pome, stone, and citrus fruits as well as to walnuts, hickories, and pecans.

Shield Budding

Use shield budding to propagate stone fruit and nut trees. On a twig of the chosen variety, cut the leaf near the base of its petiole,

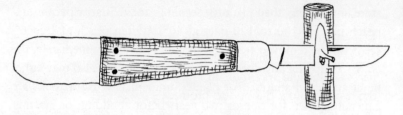

Figure 14a. Obtaining a shield bud. (Note that the stem is upside-down and that the cut is made from below.)

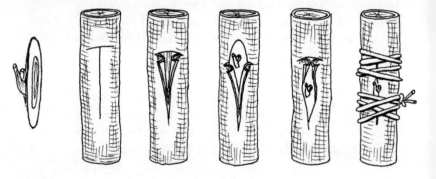

Figure 14b. The sequence of steps in the insertion of a shield bud in rootstock.

leaving only a short stump and the desired bud in the leaf axil. Then, from *below*, cut a shield-shaped portion of the bark and cambium, bearing the bud and the petiole stump (Fig. 14a). Then make a T-shaped slice in the rootstock as shown. Insert the bud from above, tie, and wax (Fig. 14b).

Flute, Patch, or Veneer Budding

Remove a square piece of bark containing the desired bud and an identical piece from the stock (Fig. 15a). Place the bud in the hole created in the bark of the stock, then tie and wax.

H-budding

Make two parallel incisions in the stock and a horizontal central cut to join them. Raise the flaps and insert the bud. Replace the flaps, and tie and wax the bud in place (Fig. 15b). You may notch the flaps to allow the bud to protrude easily.

Ring or Annular Budding

This method is similar to flute budding except that the patch is removed all the way around the stem (Fig. 15c).

Spur Budding

This method is identical to shield budding except that, instead of a bud, a short twig (the *spur*) containing more than one bud is left on the shield (Fig. 15d).

Plate Budding

This type closely resembles the H-budding method but is simpler (Fig. 15e).

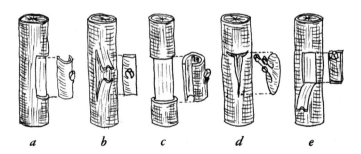

Figure 15. Other methods of budding: a. Patch, flute, or veneer budding. b. H-budding. c. Ring or annular budding. d. Spur budding. e. Plate budding.

GRAFTING

Grafting is usually performed in late winter to early spring while both rootstock and scion are dormant. In late fall cut scions from the well-matured wood of last season's growth, and keep them dormant in moist, cold sand in the root cellar until you're ready to use them the next spring. Use water sprouts, too, if they are taken from above the old graft union of the parent tree (the water sprouts from the old rootstock, below the graft union, are not genetically true). As in budding, make certain the cambium, xylem, and phloem layers closely match, making a mechanically sound union—then tie the two parts securely and wax thoroughly. Grafting should be performed under cool, moist conditions and is especially applicable to the pome fruits (for further details, see also Cox 1974b, Talbert 1946, Bailey and Bailey 1976, and Anonymous 1978b).

Whip Grafting (Whip and Tongue, or Root Grafting)

This method produces several trees from a single rootstock. Scions are fitted to root pieces taken from a single seedling. Make cuts as shown in Figure 16—a double tongue and groove—join pieces, tie, and wax. Graft in January or February and store grafts in cool, moist sand. Plant them as early as possible in the spring. Plants may bear within four to six years.

Cleft or Top Grafting

This type of grafting is best performed just after growth starts in spring. A grafting knife is especially useful for this purpose (Fig. 17a). Split the stock and insert two or more wedge-shaped scions with one edge of the cambium layers matching (Figs. 17b–d). Tie and wax (Fig. 17e). After one or two seasons, remove all but the most hardy graft. It is possible to top-graft onto trees of any age, to intergraft by this method, and to derive fruitful branches in only two to three years.

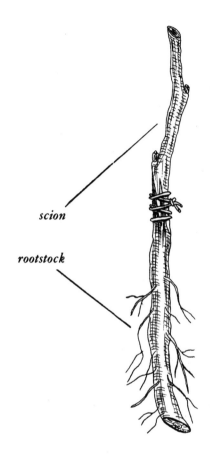

Figure 16. Scion whip-grafted to rootstock.

Bridge Grafting

When a tree is girdled, in whole or in part, the bridge graft prevents its death. This method will work on any fruit tree that can be propagated by grafting. Scions may be taken from the top of the same tree or from a close relative. Trim the wound, removing any macerated layers, then cut and insert scions as shown (Fig. 18).

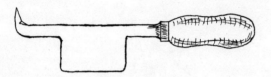

Figure 17a. A grafting knife is useful for cleft grafting.

Figure 17b. Scion prepared for insertion into cleft of stock.

Figure 17c. Two scions inserted into cleft.

ASEXUAL METHODS: BUDDING AND GRAFTING

Figure 17d. Schematic diagram of cleft graft from above, showing scions with cambium layers matching cambium layer in stock.

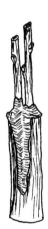

Figure 17e. Cleft graft complete and waxed.

Figure 18. *Bridge-grafting a girdled tree.*

Bark Grafting

For this method to be successful, bark should separate easily from the tree. Therefore perform bark grafting in autumn just before the leaves fall, or in spring just after the buds break. Cut the scion as shown (Fig. 19), make slits in the bark of the stock, and insert

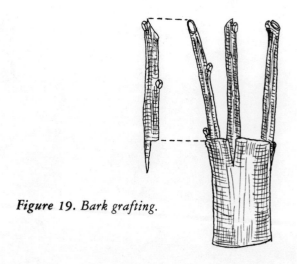

Figure 19. *Bark grafting.*

the scions so that the cambium layers are joined. Tie and wax. As with cleft grafting, remove all but the most vigorous scion after one or two seasons of growth.

Spliced and Saddle Grafting

These two grafting methods are simplifications of the whip graft except that a scion is grafted upon a stem of the seedling instead of directly to a root. The spliced graft (Fig. 20) and saddle graft (Fig. 21) are used for intergrafting different varieties on the same tree. The saddle graft may be employed when a scion is cut with a terminal bud.

In all grafting methods, any string used to tie the graft should be removed after healing occurs, or it may girdle the new graft and kill it.

Figure 20. The spliced graft.

Figure 21. The saddle graft.

Selecting the Parent Plant and Other Tips on Plant Propagation

Many factors are important in selecting the parent plant, regardless of the propagation method you use. Do not pick a plant at random; in fact, undesirable plants, especially cross-pollinators, should be pulled or eliminated *(rogued)*. Base your selection on the overall quality of the parent, taking into consideration the rate of germination, the growth habit and health of the leaves, the size, shape, and color of the fruits, their uniformity, the heaviness of the fruit set, earliness or lateness in bolting (for annuals such as spinach and lettuce), disease and insect resistance, and the quality of the root crop, among other factors. It is especially important to select parent plants carefully when producing seed. Variability is inherent in sexual methods. Asexual methods of propagation, however, almost always produce true offspring.

To propagate any plant, you must know:

1. Can the plant produce "true" seed? If so, is it perfect, monoecious, or dioecious? (Do I need more than one plant? Do I need plants of both sexes?) Also, is the plant self-pollinated or cross-pollinated, by the wind or by insects? (Do I need to separate this plant from other varieties, or even other species, in either time or space? If so, how far?) Is the plant annual, biennial, or perennial? (Can I get seed the first year? Or must I overwinter it in some way? Will I need to wait more than two years?)
2. If the plant cannot produce "true" seed, what asexual methods of propagation are available? (Is the plant herbaceous, producing bulbs, tubers, or runners? Can I take green cuttings? Is layering possible? Is the plant woody, requiring grafting or budding if hardwood cuttings won't work? Can I divide its crown or roots? Does it send up suckers or send out stolons? Will it tip-root?)
3 To what family does this plant belong? What are some of its close relatives? (Plants of the same family often behave similarly.)

Finding the answers to these questions will enable you to propagate almost any food plant of your choice. And, although I have omitted seasoning herbs and spices, cultivated ornamentals, native wildflowers, shrubs, and trees, when you have found answers to these questions you will be able to propagate these plants too. (Parsley and dill, for example, belong to the carrot family and behave just like the carrot except that dill is an annual and parsley is a biennial).

I hope this summary of some of the methods of plant propagation will encourage and enable you to become a completely self-sufficient, food-plant propagator. Remember, green plants are the basis for all life on earth and are to be known, used, understood, loved, and propagated.

PART TWO

PROPAGATING SPECIFIC FOOD PLANTS

Listed on the following pages, in alphabetical order,* are a variety of common food plants from around the world along with essential information for each plant: the country of origin, the growth form, the parts used for food, the pollination mechanism (important in seed production), seed viability (not given for plants that cannot be propagated by seed), zones where the plant is best cultivated—see map, page 41—and specific methods of propagation applicable to that particular plant. This guide should enable the conscientious gardener to propagate any of these plants successfully year after year.

*Food plants are listed in alphabetical order by common name. If the name is modified by an adjective, such as "Jerusalem Artichoke" or "Upland Cress," it appears under the basic name—in these examples, "Artichoke" or "Cress." Synonyms for certain plants, such as "Girasole" for "Jerusalem Artichoke," are listed in the index and in the Botanical Classification of Common Food Plants on page 137.

ALMOND
Prunus dulcis

ALMOND
Prunus dulcis-Rosaceae

The almond is a perennial deciduous woody tree cultivated for its edible seed (nut). The flowers are cross-pollinated by insects, and the resultant fruit resembles a shriveled apricot. The viability of the pits is 4 years. The almond probably originated in western Asia.

The almond is hardy to zones 5–8.

Propagation
Budding and Grafting
Use any conventional method to graft or bud the desired variety upon rootstock grown from viable almond pits.

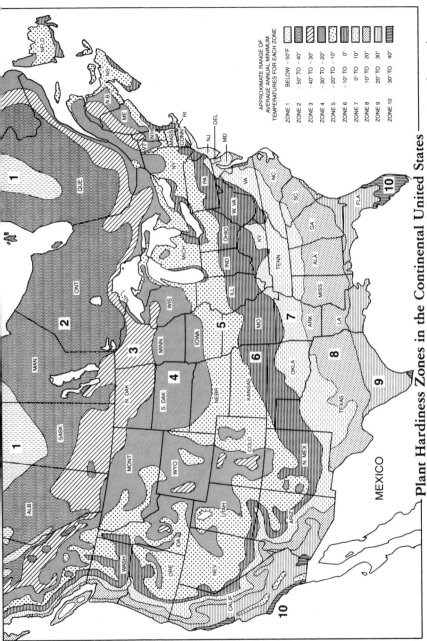

Plant Hardiness Zones in the Continental United States

These zones are generalized climatic regions that range from severe winters and short growing seasons (zone 1) to a year-round, frost-free, tropical climate (zone 10). Pockets of variation (micro-habitats) within any given zone may extend or curtail the expected growing range of some plants.

APPLE
Malus spp.–Rosaceae

The apple is a long-lived, perennial, deciduous woody tree derived from the single wild ancestor *Malus pumila*, native to southeastern Europe and southwestern Asia. The apple tree is cultivated everywhere for its large fruit. The flowers are cross-pollinated by insects in the spring, and the seeds of the apple have a viability of about 4 years.

The apple is hardy to zones 4–7, but special warm-climate varieties do well in zones 8 and 9.

Propagation
Grafting
Whip-graft a desirable variety on hardy wild apple seedling rootstock or to seedlings grown from crab apple seeds.

Budding
You can also bud a desired variety upon the same rootstock, using any method of budding.

APRICOT
Prunus armeniaca–Rosaceae

The apricot has been cultivated since 3000 B.C. It probably originated in China or Armenia. The perennial, deciduous, woody tree yields edible fruit following insect cross-pollination in the early spring. Viability of seed is about 2 years.

Apricots are hardy to zones 5–8.

Propagation
Budding and Grafting
Use any conventional method to bud or graft the desired variety upon rootstock grown from stratified apricot pits.

GLOBE ARTICHOKE
Cynara scolymus–Compositae

The globe artichoke is a robust herbaceous perennial from the Mediterranean region. The buds of the inflorescence, which are comparatively large, are eaten. If allowed to bloom, the buds open and produce thistlelike inflorescences, which are cross-pollinated by insects. Seeds remain viable for about 2 years.

This plant is hardy to zones 8–10 and does especially well in Mediterranean climates.

Propagation
Suckers
Leaving intact 5 or 6 of the most robust suckers from parent rootstock, trim the weaker ones away. Allow suckers to develop

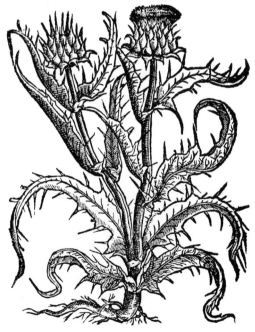

GLOBE ARTICHOKE
Cynara scolymus

to 12–18″ tall, then carefully remove them, saving as much of the root system as possible (see Fig. 6a). Make your selection from your plants of highest productivity. You will get buds the first year, but the plants produce best during the second to fourth years. Renew plants every 4 years.

Crown Division
In the fall divide the old crown into pieces including some stem and some root. This method provides more reserve food to the new transplant the next spring than does the sucker method.

Seed
Let the inflorescence bud develop on the most desirable plant. Seed will not be completely true, but it can be saved when the head becomes dry and before seed dispersal occurs, as in most plants of the sunflower family (Riotte 1973b).

JERUSALEM ARTICHOKE OR GIRASOLE
Helianthus tuberosus–Compositae

The Jerusalem artichoke is a native of North America and was grown by the American Indians. It is a tall, herbaceous perennial resembling the sunflower. Insect-pollinated flowers appear in the very late fall but are of little importance. The edible tubers, rich in inulin, serve as the exclusive means of propagation.

The Jerusalem artichoke is hardy to all zones.

Propagation
Tubers
Plant the new crop of tubers right at harvest time in the fall, taking them from the current crop. Although fall planting will give the best results in next year's growth, you may also overwinter the tubers in the root cellar and plant in the early spring.

ASPARAGUS
Asparagus officinalis–Liliaceae

Asparagus is a hardy, herbaceous perennial that dies back to its roots each fall. It originated in coastal Eurasia, where it has been cultivated and gathered wild for at least 2,000 years. In the early spring the tender young shoots are cut and eaten.

The plants are dioecious, the male plants bearing yellowish-green flowers and the female plants bearing tiny, inconspicuous flowers. Following cross-pollination by insects, small red berries develop in the fall on female plants. Seed viability will remain high as long as 3 to 5 years.

Asparagus is hardy to all zones.

Propagation
Seed
Collect the red berries from two-year-old or older female plants, harvesting before the first frost. Crush the berries and separate the seeds by hand (the seeds are large, shiny, and black) or by immersing them in water. The pulp will float as the seeds sink. Dry the seeds for 2 to 3 weeks. In the spring grow the new plants in deep, loose soil. After one season transplant them to a permanent bed. Transplant in the early spring before growth resumes or in the fall after growth has ceased. If plants are dug in the fall, the roots should be stored in the root cellar at 40°F until planting time.

Root Division
Asparagus roots may be divided and replanted. Before picking shoots, allow one year for the plants to reestablish good root systems. Then pick sparingly during the second year.

AVOCADO
Persea americana–Lauraceae

The avocado is a tropical or subtropical American tree bearing soft-textured fruit. This perennial, broad-leaved evergreen grows well only in frost-free areas. The flowers are cross-pollinated by insects. The viability of the large seed is one year.

The avocado can be grown in zones 9–10.

Propagation
Shield Budding
This is the best method to join a desired variety to seedling stock grown from the avocado seed. Sprout the seed in water or moist sand; the seed may require 6 to 8 weeks to germinate. Then transplant it to a pot. After the graft has taken, place the tree in a permanent outdoor location.

Cleft Grafting
You can also cleft-graft a desired variety to the seedling stock. Both budding and grafting may be done in early spring or late fall.

BANANA
Musa paradisiaca–Musaceae

The banana, one of the oldest known human foods, comes from India and China. Although it is herbaceous and perennial, it is a tree and is cultivated for its fruit. The clustered flowers are pollinated by bats; however, the banana rarely produces seed and can be propagated only vegetatively.

It is grown in zones 9–10 and can withstand only a few degrees of frost.

Propagation
Root Cuttings
Plant pieces of the rhizome, 3 to 4 pounds each, and each with a bud or "eye." Cover completely with soil; each will form a new plant.

Suckers
Suckers arise from the base of the exhausted "trunk" of the parent plant, which fruits only once, and may be removed, roots intact, to be planted elsewhere.

BARLEY
Hordeum vulgare–Gramineae

Barley, of unknown nativity, has been grown since the dawn of civilization. The plant is thought to have originated somewhere in the Old World. It is a herbaceous annual grass, its seed used for human food or animal food, or sprouted for malt. Flowers are tiny and inconspicuous and are cross-pollinated by the wind. Seeds remain viable for at least one year, but not much longer.

Barley can be grown in zones 3–8.

Propagation
Seed
Cut mature, dry stalks carefully, and thresh the seed from the heads, or spikes. Then winnow out most of the chaff and remaining debris. Dry the seed further and store. Plant in the early spring.

BEAN
Phaseolus spp., *Vigna spp.*, and *Vicia faba*–Leguminosae

Most types of bean are annual herbs cultivated for their seeds or pods. They are found in all parts of the world. The adzuki bean

KIDNEY BEAN
Phaseolus vulgaris

(Vigna angularis) is from Asia. Broad beans *(Vicia faba)* have been known in Eurasia for 3,000 years, are very hardy, and yield snap, shell, and dry beans. The kidney bean *(Phaseolus vulgaris)* appears to be of ancient South American origin and can be found in many varieties as twining or bush annuals. Kidney beans yield snap, shell, and dry beans. The lima bean *(Phaseolus limensis),* tender, twining, or bushy, comes from tropical America (Guatemala) and also yields shell or dry beans. The moth bean *(Vigna aconitifolia)* is from East India. The scarlet runner or multiflora bean *(Phaseolus coccineus),* from Central or South America, is the only perennial in the group, but it too is grown as an annual, in

twining and in bush form. Young pods or green or dry seeds are eaten. The mung bean *(Vigna radiata)*, a tender plant from East India, is grown mostly for its sprouts. The tepary bean *(Phaseolus acutifolius)* was domesticated by prehistoric Indians of the southwestern United States and northern Mexico. This plant can withstand extreme heat and drought. The urd bean *(Vigna mungo)* comes from East India. All beans are generally self-pollinating, requiring no external pollinator. Seed viability is 3 years.

Plant beans in zones 3–10 for the *Phaseolus* species, which are warm-season crops sensitive to frost, in zones 3–6 for *Vicia faba*, which do especially well in maritime climates. (Broad beans dislike hot, dry summers.)

Propagation
Seed
Every variety of bean may be propagated in the same manner. Allow the pods to dry on the vine in the fall. Pick them before the first frost, and shell the beans from the pods. Dry the beans for another month and store them to overwinter. It is also possible to pick the yellow pods and dry them for 2 weeks; then shell them and treat as above.

The only exception to the above is the mung bean. In this case, pull mature plants with drying pods. Dry the entire plant in windrows and then thresh them. The pods shatter easily, and the seeds will then fall out. Collect the seeds, dry them for 2 to 3 weeks, and store until spring.

RED BEET, GARDEN BEET
Beta vulgaris–Crassa Group–Chenopodiaceae

Beets are herbaceous biennials from Europe, northern Africa, and western Asia. They are cultivated for their large spherical roots and leafy greens. The small, densely clustered flowers are cross-pollinated by the wind. The seeds will remain viable for 5 to 6 years.

Beets are hardy to all zones.

Propagation
Seed

First-year beets form a thickened root and a rosette of leaves. Save the best beets, which will be uniformly red, to overwinter in the root cellar or in the ground, if they are protected from severe freezing. In the early spring, replant the beet root in the desired location, and stake the elongating flowering stems so they will not collapse. When the seed near the bottom of the stalk is mature, pull up the entire plant and hang it in a warm, dry place or dry it on a canvas sheet. When the plant is completely dry, rub the flower stalk between your hands to break the seeds from the stalk. Screen out the seeds from the other debris. Dry the seeds and store them to overwinter.

Beet varieties will cross, and beets of any variety will cross with sugar beets, mangels, or Swiss chard. Therefore grow only one of these varieties at a time, or place them far enough apart that the wind cannot carry the pollen. A barrier such as a building will also block pollen transfer effectively.

SUGAR BEET
Beta vulgaris–Crassa Group–Chenopodiaceae

Sugar beets are a form of red beet except that they have a larger and longer root and no red pigmentation. The sugar beet most likely originated in Germany and, like the red beet, is a herbaceous biennial. Its large root is used as feed for livestock or for its sugar. Flowering and seed production are exactly the same as in the red beet; the inconspicuous flower clusters are cross-pollinated by the wind. Seed viability is 3 to 5 years.

The sugar beet grows in all zones.

Propagation
Seed
Treat the sugar beet exactly as you would the red beet. Also note the possible undesirable crosses.

BLACKBERRY
Rubus spp.–Rosaceae

The blackberry is a North American perennial woody vine with biennial canes. The plant is armed with spines and tends to arch and climb. The flowers are cross-pollinated by insects and yield sweet, juicy berries. The blackberry is propagated totally by vegetative means.

The blackberry is hardy to zones 3–9.

Propagation
Root Cuttings
Choose roots the thickness of a lead pencil. Cut each root into 3–4" lengths, and bury them in the soil to induce callus growth over the winter. In the spring, plant the cuttings 3–4" apart and 3" deep. One year later, transplant the young plants to their permanent site.

Suckers
Cut the nonfruiting, current year's suckers away from the parent plant, roots attached, in the fall after the leaves have dropped. Transplant them to the desired site. The new plants will fruit the following spring.

Tip Rooting
Trailing or lopping varieties of blackberry root at the tips where the canes touch the ground (see Fig. 10). New plants may be dug up, separated from the parent plant, and moved to a new location.

BILBERRY, WHORTLEBERRY
Vaccinium myrtillus

BLUEBERRY
Vaccinium spp.–Ericaceae

The blueberry is a perennial, woody shrub native to North America. Its smooth, blue, globose berries are eaten. Its small, vaselike flowers are cross-pollinated by insects. Seeds remain viable for up to 3 years.

Blueberries grow in zones 2–9.

Propagation
Root Division
In the winter, divide the dormant rootstock into several pieces, each retaining a portion of the crown.

Mound Layering

Cut back the parent plant, and wound the base of each upright branch before mound layering (Fig. 9).

Stem Cuttings

Make cuttings after freezing weather has set in to be sure that the plants are hardened and dormant. Overwinter the cuttings in moist sand. In the spring, root each cutting in potting soil. After one year, transplant the cutting to its permanent site.

Seed

Seeds can be separated from mature fruits by crushing the berries and floating them in water. Good blueberry offspring, while not exactly true, may be produced from these seeds.

SPROUTING BROCCOLI
Brassica oleracea-Italica Group-Cruciferae

Sprouting broccoli, also called branching or asparagus broccoli, is the ordinary plant we find in the supermarket. It is a herbaceous annual from the coast of western Europe. Its green flower-bud shoots are eaten. The flowers are cross-pollinated by insects, and the seeds will remain viable for 5 years.

A cool-season crop, broccoli grows well in zones 3-8.

Propagation
Seed

Allow one or more flower-bud shoots to elongate and bloom fully. A raceme of yellow flowers will form, to be visited by bees and butterflies. The flowers mature into elongated siliques, resembling numerous pods on the stem. Pull the entire plant when the oldest siliques are mature and are beginning to split. Hang the plant in a warm, dry place (this continues the development of the seeds on the drying parent plant) until thoroughly dry; then rub the dry siliques through a small screen to separate the small, round seeds. Dry the seeds for an additional 2 weeks, and store until spring.

BRUSSELS SPROUTS
Brassica oleracea–Gemmifera Group–Cruciferae

Brussels sprouts are a herbaceous biennial that originated in Europe, perhaps in Belgium. Globular leafy sprouts, resembling tiny cabbages, appear in the leaf axils of the lower stem and are eaten as a vegetable. The flowers are cross-pollinated by insects, and the seeds will remain viable for 4 to 5 years.

Brussels sprouts are a cool-season crop that does well in zones 3–8.

Propagation
Seeds
Mulch the plant heavily to protect it through the winter in severe climates; in milder climates, simply leave the plant in the garden. Elongated flower stalks will form during the second season of growth. Harvest the plant when the oldest siliques are yellow but not dry (see instructions for sprouting broccoli). Since the seed pods burst when dry, the seeds are easily separated. Collect them on canvas or in paper bags, making certain that they are dry before storing.

BUCKWHEAT
Fagopyrum esculentum–Polygonaceae

Buckwheat is a herbaceous annual from Manchuria, cultivated either for its seeds, which are ground into flour, or simply as a bee plant for honey production. The flowers are cross-pollinated by various insects. Seed viability is high for one year.

Buckwheat does well in all zones.

Propagation
Seed

Buckwheat seeds form in spikelike clusters at the top of the plant. The seeds have triangular wings. Cut the mature stalk of seeds, and thresh and winnow out foreign matter and broken wings. Dry further and store the seeds for the winter, either to grind into flour or to plant the following spring.

BUTTERNUT
Juglans cinerea–Juglandaceae

Butternut is a perennial, deciduous tree found in eastern North America from New Brunswick to Arkansas. The nuts have long been popular among Indians and settlers. However, the tree is less abundant today than it once was because many butternuts have been cut for lumber. The trees are monoecious, and the catkins wind-pollinated. The female flowers develop into clusters of nuts at the branch tips.

The butternut grows in zones 4–7.

Propagation
Grafting

Graft the scion from a productive tree to seedling stock grown from stratified butternut or walnut seed. You may also intergraft the butternut with any of the other walnuts or any species of *Juglans*.

HEAD CABBAGE
Brassica oleracea–Capitata Group–Cruciferae

Head cabbage originated on the coastal chalk rocks of southeastern Europe and has been cultivated since 2500 B.C. It is a herbaceous biennial with a dense, globular, leafy head eaten as greens. The flowers are cross-pollinated by insects; however, the plants are incompatible, or self-sterile. Seed production, then, will depend upon two or more cabbage plants crossing. Head cabbage seed will remain viable for 5 years.

Hardy to all zones, head cabbage is a cool-season crop that will withstand periods of considerable frost.

Propagation
Seed

Grow the head the first season, and overwinter it in the root cellar at a temperature as near to 32°F as possible. Store the head in moist peat, or mulch the plant heavily and leave it in the garden if winter temperatures do not fall below 25°F. In the spring, make two cuts at right angles on the top of the head to allow the seed stalk to grow normally—otherwise it may curl within the leaves. After the seed stalk flowers, remove the seeds, following the instructions for sprouting broccoli.

Varieties of head cabbage will cross with each other, with wild radish and with wild mustard, so it is best to grow only one variety a year for seed. A single head is unable to produce seed, so grow at least two heads of one variety.

PAK CHOI CABBAGE
Brassica rapa–Chinensis Group–Cruciferae

Pak choi (bok choy) cabbage is a herbaceous biennial or early spring annual from China. The leaves are eaten as a potherb. The flowers are cross-pollinated by insects and the resulting seeds are viable for 3 to 5 years.

The pak choi is a cool-season crop hardy to zones 3–8.

Propagation
Seed
When this cabbage is grown as a late summer biennial, its seed production is similar to that of head cabbage (see head cabbage propagation). If planted very early, pak choi behaves like an annual, producing seed the first year. The flowers, fruits, and seeds closely resemble those of broccoli or cabbage and are collected in the same manner.

PE-TSAI CABBAGE
Brassica rapa–Chinensis Group–Cruciferae

Pe-tsai cabbage is also from China and has been cultivated since 2000 B.C. It is a herbaceous annual whose leaves serve as a potherb or are used raw in salad. Seeds develop from insect-pollinated flowers and will remain viable for 5 years.

Pe-tsai is hardy to zones 3–8.

Propagation
Seed
See instructions for sprouting broccoli or head cabbage.

CANTALOUPE
Cucumis melo–Cantalupensis Group–Cucurbitaceae

Cantaloupe, like muskmelon, is a herbaceous annual vine with a tendency to sprawl. It originated in southern Asia. Its large, succulent fruit is eaten. The plants are monecious, female flowers appearing after the male flowers. Cross-pollination occurs by insects, and seeds will remain viable for 5 years.

A warm-season crop, the cantaloupe is found in zones 5–10.

Propagation
Seed
Select fruit from vines bearing several uniformly shaped cantaloupes with good netting. A relatively heavy weight for the size indicates a good-quality fruit with a small seed cavity. Follow instructions for muskmelon seed propagation.

CARDOON
Cynara cardunculus–Compositae

Cardoon is a herbaceous perennial originally from southern Europe. Its root and leaves are used for food. The flowers are cross-pollinated by insects.

Cardoon is hardy to zones 8–10.

Propagation
Seed
See instructions for endive and chicory.

Suckers
Suckers arise at the base of the parent plant, much as they do on the globe artichoke, and are easily separated. Cut out the suckers when they are 8–10″ tall, and include some of the roots. Transplant the sucker immediately to a new location.

CARROT
Daucus carota var. *sativus*–Umbelliferae

The carrot is a herbaceous biennial descended from the wild carrot of Eurasia and northern Africa. Its taproot is edible. Dense com-

CARROT, BIRD'S NEST
Daucus carota

pound umbels of tiny white flowers grow on a tall stalk and are cross-pollinated by insects during the second season. Carrot seeds remain viable for 2 to 3 years.

Carrots are hardy to all zones.

Propagation
Seed

Grow carrot roots the first year. Save roots that are uniform in size, shape, and color and that have a small core. Overwinter these carrots in the root cellar, or leave them in the garden if there is no danger of extreme cold. In severe climates, a heavy mulch will prevent the carrots from freezing. In the early spring, plant the overwintered carrots in a row 3' apart, leaving at least 1" of root below the uninjured crown. Later in the summer, when the secondary heads are ripe and the tertiary heads are browning, pull the entire plant and hang it upside-down for 1 to 2 weeks in a warm, dry room or attic. Then cut off the dry clusters of seeds, and crumble them between your hands. Use a winnow or sieve to remove the stem debris. Dry the seeds a few more days and store.

Cultivated carrots will cross readily with the wild carrot, Queen Anne's lace, because they are of the same species. The domestic carrot is a descendant of the wild carrot, which also yields a tasty, if smaller and paler, taproot. Because of this possible crossing, which would affect the quality of the next generation, be careful to isolate your carrot plants from their wild ancestors whenever you grow them for seed.

CAULIFLOWER AND BROCCOLI
Brassica oleracea-Botrytis Group-Cruciferae

Cauliflower and broccoli are herbaceous biennials from coastal Europe. Cauliflower is cultivated for its curd, the white, tender flower-bud clusters. Broccoli in the Botrytis Group (not to be confused with sprouting broccoli on page 53) is closely related to the white cauliflower and is similar to it except that the curd is green. The flowers that appear during the second season of growth are cross-pollinated by insects, and the seeds remain viable for 5 years.

The cauliflower is grown in zones 4–7 and requires cool, moist, stable growing conditions.

Propagation
Seed

Save an old plant from the current season; you may even pick the curd. After overwintering the plant in the root cellar or in the field, set the plant outdoors in the desired location. Flowering and seed-bearing shoots will come from the stump and any parts of the head that have not been removed. After the flowering proceeds, treat cauliflower according to the instructions for cabbage; they are really varieties of the same species.

Cauliflower is more difficult to grow than cabbage, and it does best in cool, moist situations. Strict selection of parent plants will ensure the quality of succeeding generations.

CELERIAC
Apium graveolens var. *rapaceum*–Umbelliferae

Although the origin of celeriac is obscure, this herbaceous biennial has been cultivated since the seventeenth century for its tuberous base or root crown. The flowers are cross-pollinated by insects, and the seeds remain viable for 5 years.

Celeriac does well in all zones and prefers rich, moist soil.

Propagation
Seed
Treat celeriac as you would carrot or celery.

CELERY
Apium graveolens var. *dulce*–Umbelliferae

Celery, like celeriac, is a herbaceous biennial. The plant originated in marshy places ranging from Sweden to northern Africa to eastern Asia. The compound umbels of tiny flowers are cross-pollinated by insects; seed viability is 5 years.

Cultivated for its fleshy and fibrous leaf petiole, celery does well in all zones and prefers rich, moist soil.

Propagation
Seed
Store the selected parent plant grown the first season in the root cellar or in the garden over the winter, but protect it from severe freezing. Set the plants out the following spring. After the flowers and seeds form, treat celery according to the instructions for saving carrot seed.

SWISS CHARD
Beta vulgaris–Cicla Group–Chenopodiaceae

Swiss chard comes from the Canary Islands. This herbaceous biennial yields large, fleshy leaf stalks that make a tender potherb. The flowers are tiny, densely clustered on the stalk, and cross-pollinated by the wind. The seeds remain viable for 4 to 6 years.

Swiss chard does well in all zones.

Propagation
Seed
Leave the parent plant in the garden the first year, or transplant it to a pot to be overwintered where it is protected from severe freezing. Replant it the following spring; the seed stalk will form as in the red beet. Stake the flowering stalk as necessary to keep it from

falling. Harvest the entire plant when the seed at the base of the stalk is mature and brown. Hang the plant in a warm, dry place to dry completely; then treat it following the instructions for red beet propagation.

Swiss chard will cross with beets (they are varieties of the same species) so grow only one of these for seed each season, alternating seed-saving years. Grow enough seed to last for two or more years when you grow either plant.

CHAYOTE OR VEGETABLE PEAR
Sechium edule-Cucurbitaceae

The chayote is a perennial-rooted vine of Mexico, Central America, and the West Indies and is often grown as an annual. Its fruits and tuberous roots are edible. The plants are incompatible, or self-sterile, and cross-pollination by insects must occur between two different plants. Seeds remain viable for only one year.

The chayote is mainly a tropical plant but is adaptable to southern coastal regions of the United States, from South Carolina to southern California; it is cultivated in zones 9–10.

Propagation
Seed
Save a mature fruit from each desired parent. Plant the entire fruit (one seed per fruit) in the spring with the broad end sloping downward, leaving the stem end slightly exposed. To produce seed, you will need two parent plants.

Cuttings
Special varieties of chayote may be propagated by shoot cuttings or green cuttings taken at the crown of the plant. Root these in loose soil in a greenhouse pot (Riotte 1974).

CHERRY

Prunus avium (sweet) and *Prunus cerasus* (sour)–Rosaceae

The cherry is of obscure origin but is thought to come from the area of the Black and Caspian seas. The perennial deciduous tree yields small stone fruits in early summer. The flowers are cross-pollinated by insects; the seeds within the pits are viable for one year.

The cherry grows well in zones 3–10.

Propagation
Budding

Use *Mazzard* (*Prunus avium*) or *Mahaleb* (*Prunus cerasus*) rootstock, or grow your own from cherry seeds taken from the fruits and planted immediately (drying will kill the embryos). You can enhance germination by gently cracking the bony shell of each pit and removing the seed. Plant the seeds in flats or individual pots, and then stratify them (see instructions under "Stratification" on page 12) for 110 to 120 days at 40°F. Following the period of stratification, set the seedlings out in the nursery to develop for 1 to 2 years before budding.

CHERVIL

Anthriscus cerefolium–Umbelliferae

The chervil, a hardy herbaceous annual originally from Europe, furnishes leaves that are used in salads or as garnishes. Flowers are produced in compound umbels and are cross-pollinated by insects. Seed viability is 2 to 3 years.

Chervil grows in all zones.

Propagation
Seed

The chervil sends up seedstalk that resembles that of the carrot the first year. Treat this plant in the same way as the carrot and

CHERRY
Prunus cerasus

most umbellifers. Pull the entire plant when the seeds are mature, and hang it in a warm dry place to finish developing and to dry fully. Then rub the seeds from the dry heads, winnow or screen, and store until spring.

CHESTNUT
Castanea spp.–Fagaceae

The American chestnut *(C. dentata)* is a deciduous tree native to the forests of eastern North America. The nuts of the chestnut were eaten by Indians and settlers, much the same as the butternut fruit, but the trees have become scarce due to lumbering and Dutch elm disease.

A few American chestnuts still remain and yield chestnuts. It may be more practial, however, to cultivate two other species of chestnut that are disease-resistant, the Japanese chestnut *(C. Crenata)* and the Chinese chestnut *(C. mollissima)* in areas of the eastern United States where the native American chestnut has been affected by the disease. All chestnuts are monoecious and bear unisexual catkins, which are wind-pollinated. The female flowers develop spiny fruits, which split at maturity to yield 2 to 3 nuts.

The tree is hardy to zones 4–8.

Propagation
Budding and Grafting
Use native seedling rootstock for native varieties. Expert workmanship is required to graft *Castanea* species successfully. It is advisable to develop skill with easier grafts before attempting to work with this species. Once the skill is developed, it is possible to intergraft the several species of chestnut—the Chinese, Japanese, and Eurasian forms of *Castanea*.

CHICORY
Cichorium intybus–Compositae

The chicory is a herbaceous perennial originally from Eurasia but now commonly naturalized along roadsides throughout North America. Its leaves are used in salads, and the root may be dug and roasted to make a substitute for coffee. The large, bluish flowers are cross-pollinated by summer insects. The seeds will remain viable for 4 to 8 years.

Chicory grows well in zones 3–9.

Propagation
Seed
Collect the seed from mature heads after they turn fuzzy. Separate the seed from the "chaff" by rubbing it through your hands and winnowing.

Root Division
In the late fall, after the plants become dormant, dig up and separate the rootstock, leaving at least one perennating bud on each section to be transplanted. Plant immediately in the new site, or overwinter in the root cellar and plant in the spring.

CHIVE
Allium schoenoprasum–Amaryllidaceae

Chives are tufted herbaceous perennials native to Eurasia. The rounded and hollow leaves are used as a seasoning. Flowers are cross-pollinated by insects, and the seeds are viable for 2 years.

Chive grows in all zones.

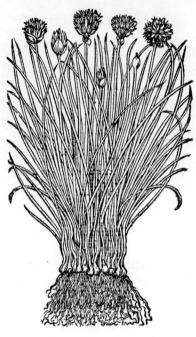

CHIVE
Allium schoenoprasum

Propagation
Seed

In midsummer, purple flowers form in globular heads. Let them develop into clusters of splitting fruits. Black seeds can be shaken out when ripe (blackness indicates ripeness) and dried and stored, to be planted the following spring.

Clump Division

The tufted clump may be dug and split into several sections, each of which, when transplanted, will grow rapidly into a new clump.

COCONUT
Cocos nucifera–Palmae

The coconut palm tree is of Asiatic or Polynesian origin but has been dispersed widely throughout tropical and subtropical islands and coastlines. The fruit is a large nut with a thick inner layer and liquid "milk," both of which are edible and tasty. The flowers of this perennial are cross-pollinated by insects.

The coconut grows in tropical climates and is hardy to zone 10.

Propagation
Seed
Simply take an untreated coconut, which is actually a single seed, and plant it in a shaded seedbed with the nut not completely covered. After germination and seedling growth, transplant the plant once or more before placing it in a permanent location. The young tree will not produce a maximum yield until it is about 10 years old.

COLLARDS
Brassica oleracea–Acephala Group–Cruciferae

Collards, like so many varieties of *Brassica oleracea*, originated in Europe. The herbaceous biennial, which is potentially perennial, provides leaves that are used as greens. The flowers, in tall racemes, are cross-pollinated by insects, and seeds remain viable for 3 to 5 years.

Collards grow in zones 3–8.

Propagation
Seed
Where winters permit, rootstocks from the current year may be allowed to stand in the garden; otherwise store them in the root

cellar. Plant the overwintered plants in the spring. During the second season, the plant will flower and produce seed in exactly the same manner as broccoli, cabbage, or any of the crucifers. (See instructions for propagating these vegetables.)

CORN
Zea mays–Gramineae

Corn is of ancient origin, and the exact nature of its development is shrouded in mystery; but it seems to have come from South America, probably Peru. All forms of corn are herbaceous annuals; all are cross-pollinated by the wind. The plants are monoecious—the "tassel" consists of male flowers, and the "ear" of female flowers. Corn silk is made up of numerous long styles, which extend from the ear and catch the pollen. Popcorn (var. *praecox*) produces seeds that explode when subjected to heat; sweet corn (var. *rugosa*) yields seeds that may either be eaten as is, removed for canning, or dried and removed to be used in various ways, especially as corn meal. Seed viability is high for 1 to 3 years.

Corn is hardy to all zones.

Propagation
Seed

Use only nonhybrid, open-pollinated varieties for true-breeding stock, such as Golden Bantam. It is important to remember that hybrid plants will not breed true. Allow the ears to mature on the stalk; harvest them after the plants have dried in the late fall before the frost. Ears may be husked, dried further, and then shelled; twist the grains off the ear by hand. After shelling, dry the corn further on screens and store. Handle the seeds gently to avoid damaging the germ. Also, for best pollination, hence seed set, plant corn in blocks of short rows rather than in long single rows. This will enable the wind to disperse the pollen effectively.

MAIZE (INDIAN CORN)
Zea mays

CRANBERRY
Vaccinium macrocarpon–Ericaceae

The cranberry is a perennial, evergreen, creeping subshrub found only in boggy areas of the boreal northern hemisphere and cultivatated for its fruit. The flowers are probably cross-pollinated by insects.

The cranberry thrives in acid bog conditions and is hardy to zones 3–6.

Propagation
Stem Cuttings
Cut stems from the parent in the early spring just before the terminal buds break, and set them immediately in their new locations.

Do not allow them to dry out. Cuttings should be 5–10" long and placed 12–18" apart in each direction. Three years later the plants will begin to bear fruit.

GARDEN CRESS
Lepidium sativum-Cruciferae

Garden cress is a herbaceous annual originally from Europe. The leaves are used for salads or as a garnish. The flowers are cross-pollinated by insects, and seeds remain viable for 5 years.

Garden cress is hardy to zones 3–8.

Propagation
Seed

During the late spring, the plant will send up an elongated flowering stem bearing numerous tiny white flowers. After pollination, small flattened fruits develop, each containing two seeds. When they begin to mature, becoming dry and brown, pull the entire plant. Hang the cress to dry further, putting a paper bag over the stem to catch the seeds that drop out as the dried fruits split. Shake the plant when it is completely dry; the seeds for next year's crop will fall cleanly into the bag.

WATERCRESS
Nasturtium officinale-Cruciferae

Watercress is a perennial herb found in aquatic places, especially slowly flowing, clear cold steams. Although originally from Great Britain, watercress is now widely naturalized throughout the cooler portions of the northern hemisphere. The leaves are used in salads, in sandwiches, and as a garnish. The tiny white flowers are cross-pollinated by insects, and the seeds remain viable for 2 to 5 years.

Watercress is hardy to zones 3–8 and prefers very wet soil.

Propagation
Seed
Watercress growing in the wild will produce a raceme of tiny white flowers, each later producing a small silique containing seeds. Pick the maturing inflorescences and hang them to dry, keeping a sack over the fruits to prevent seed loss. The seeds can simply be shaken out after the mature fruits split.

Stem Cuttings
Cut the leafy stem near the base; often you can get a bit of the root with the stem. Plant it immediately in a very moist spot or in a slow-moving nearby creek where greens may be harvested later. The green cutting will root and flourish if the site chosen is suitably moist and cool.

WINTER CRESS
Barbarea vulgaris-Cruciferae

Winter cress is a herbaceous biennial or perennial from Europe, the leaves of which are eaten. The flowers are cross-pollinated by insects, and the seeds remain viable for 2 years.

Winter cress is hardy to zones 3–8.

Propagation
Seed
Winter cress will flower during its second year, closely resembling any other crucifer in growth habit. Seeds may be collected in the same manner as for the *Brassica* species.

CUCUMBER
Cucumis sativa-Cucurbitaceae

The cucumber is an annual herbaceous vine from Asia and Africa that has been cultivated for more than 3,000 years. The fruit is

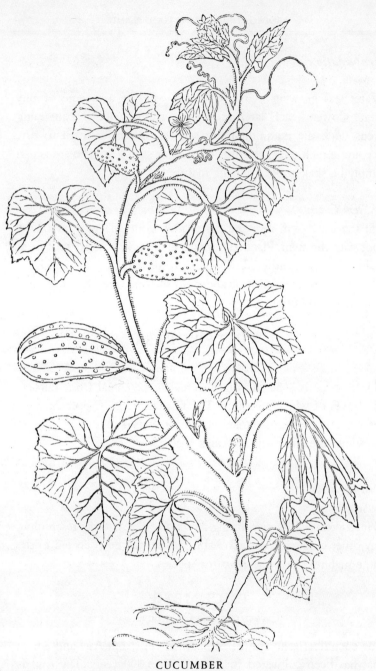

CUCUMBER
Cucumis sativa

eaten while it is still young and tender, either raw—whole or in salads—or pickled. The plants are monoecious, the male flowers appearing first and remaining more numerous than the female flowers. The flowers are cross-pollinated by insects. The seeds will remain viable for 3 to 5 years.

The cucumber is a warm-season crop hardy to zones 4–10.

Propagation
Seed

Select fruit that is long, slender, straight, and dark green; avoid fruit with whitish or yellowish streaks. Fruit produced later in the summer (second and third sets) usually produces more seed than earlier fruit. Let the chosen fruit vine-ripen until it is yellow-orange, even allowing it to become mushy. Then pick the fruit, cut it in half, scrape out the seeds, and let it ferment in a crock or wooden vessel for a few days. To assist the fermentation, add a bit of water to the seeds. Stir occasionally. The good seed will sink. Decant the liquid, wash the seed, and dry it quickly for 2 to 3 days on a screen, then store.

Cucumber varieties will cross, of course, and should not be grown together for seed. In fact, the entire *Cucumis* genus is interfertile (see Fig. 2), and members of this group should not be grown together if seed is desired.

LEMON CUCUMBER
Cucumis melo–Chito Group–Cucurbitaceae

The lemon cucumber, or mango melon, is also an annual herbaceous vine from Asia and Africa. It yields fruit that is eaten when young and tender in the same manner as cucumbers. The plants are monoecious and cross-pollinated by insects, with a seed viability of 3 to 5 years, much the same as cucumbers.

Lemon cucumber is a warm-season crop and is hardy to zones 5–10.

Propagation
Seed

Follow instructions for regular cucumber. Note, however, that the lemon cucumber is not a cucumber at all; rather it is a form of melon closely related to the muskmelon and cantaloupe, with which it readily crosses.

RED AND BLACK CURRANT
Ribes sativum and *Ribes nigrum*–Saxifragaceae

Currants are perennial woody shrubs of the cooler climates of the boreal northern hemisphere. The small globose fruit is eaten or used in jams and jellies. The flowers are cross-pollinated by insects. In general, currants are not spiny, but their close relatives, the gooseberries, are.

Currants are hardy to zones 4–6.

Propagation
Stem Cuttings

Vigorous shoots should be cut in late autumn and either planted immediately or stored in moist sand until spring and then planted. This type of cutting is called a dormant cutting (Hills 1975).

Mound Layering

This method is most successful when the plant stems are cut back to the ground before mounding, ensuring that all the shoots are only one season old.

DANDELION
Taraxacum officinale–Compositae

The dandelion is a familiar, herbaceous perennial from Europe and Asia, now of cosmopolitan distribution. The leaves make a good spring salad, either raw or cooked; the root may be dug up,

DANDELION
Taraxacum officinale

roasted, and steeped to make dandelion tea. Flower heads are cross-pollinated by insects and bloom from early spring to late fall. Seeds, blown about by the wind on tiny hairlike parachutes, remain viable for 3 years.

Dandelions are hardy to all zones.

Propagation
Seed

This plant crops up spontaneously in fields, lawns, and gardens. However, improved horticultural varieties are available for propagation. Allow the flower heads to develop the familiar "fuzz-ball" seed cluster, then pick the cluster and shake it into a paper sack to collect the seed. Run the mixture through a small screen or winnow to separate the seed from the fragments of the pappus, the hairlike parachute. Dry the seed for one day, and store until spring.

DATE
Phoenix dactylifera–Palmae

The date plam is a woody tree of ancient cultivation from Arabia or northern Africa. It is now widespread throughout the subtropical parts of the world, where it is cultivated for its small but very sugary fruit. The plants are dioecious, bearing only male or female flowers. Cross-pollination is effected by wind.

Dates require warm temperatures and do well in zones 9–10.

Propagation
Suckers

Also called "tillers," these shoots arise from the base of the parent plant and will be of the same sex. Remove the suckers when they are 3 to 6 years old. Cut them at the base, taking some root, and plant them immediately, allowing no drying period. Transplanting should be done in spring. After transplanting a sucker, cut back part of its leaves. This will reduce water loss by transpiration and allow the newly planted sucker a more rapid recovery. In five to six years, the new tree should be sexually mature and will begin to flower. When propagating by suckers, pick mostly female plants for fruit bearing and a few male plants to provide pollination. Plant the suckers strategically so that the wind will carry the pollen from the males to all of the females.

Seed
Dates should be picked and dried and the pits removed. The pits can be planted in greenhouse pots to sprout and develop into young trees. Offspring will not come completely true, and the sex cannot be known until flowering occurs. As in the case of suckers, both sexes will be needed. When propagating a named variety, do not use this method; use the sucker method instead.

DEWBERRY OR LOGANBERRY
Rubus ursinus–Rosaceae

The western dewberry is a vine or shrub native to western North America. The loganberry is a cultivated descendant of the western dewberry and a variety of the same species. All of these yield juicy aggregate fruits similar to the blackberry, which is a close relative. Their flowers are cross-pollinated by insects.

The dewberry is hardy to zones 4–8.

Propagation
Root Cuttings
See instructions for blackberry propagation.

Tip Layering
This method is actually forced tip rooting. Cover the tips of the canes before the harvest. Roots will form by the following spring.

Then, disconnect and transplant the new plants before spring growth resumes.

EGGPLANT
Solanum melongena–Solanaceae

The eggplant is a herbaceous annual of Africa and Asia, known since the fifth century A.D. The large soft fruit is edible. The flowers are self-pollinated, and seed viability is 4 to 5 years.

Eggplant requires a long season and is hardy to zones 7–10.

Propagation
Seed

One fruit will provide enough seed for several hundred plants. Choose one fruit and let it mature just past the edible stage. Cut the fruit in half, scrape out the seeds, and wash them free of pulp. Seed should be dried immediately in full sun on a screen. Store dry seeds until spring.

ENDIVE
Cichorium endivia–Compositae

Endive is a hardy herbaceous annual or biennial originating in India and known to have been used by the Egyptians. The leaves are used as a salad green or as a potherb and the flowers are cross-pollinated by insects in mid to late summer. The seed will remain viable for 4 to 5 years.

A hardy crop, endive may be grown from late summer through early winter and thrives in zones 3–9.

Propagation
Seed

See instructions on seed propagation for endive's close cousin, chicory.

FIG
Ficus carica–Moraceae

The fig tree, which comes from Syria, is a woody deciduous perennial yielding a soft fruit. Tiny wasps cross-pollinate the clusters of small flowers, which then in-roll and mature to the multiple fruit that is the fig. Methods of propagation are vegetative; seeds are not used.

Figs do well in zones 7–10.

Propagation
Cuttings
Hardwood cuttings taken in the fall should be overwintered and planted the following spring. They may be greenhouse-grown for one season and then transplanted to a permanent location.

Budding and Grafting
Most conventional methods are applicable. Stock can be grown from the tiny seeds within the fig, and desirable varieties can be grafted or budded to this stock after one year.

Suckers
Rooted suckers from the base of the tree may be transplanted (Wahlfeldt 1971).

FILBERT OR HAZELNUT
Corylus spp.–Betulaceae

Filberts and hazelnuts are known to be native to the United States, Europe, and western Asia. The bushy tree yields a hard-shelled nut in a papery enclosure. The plants are deciduous and monoecious, bearing the sexes in separate catkins. Catkins appear in very late winter or early spring, before the leaves come out, and are cross-pollinated by the wind.

Filberts are hardy to zones 4–9.

Propagation
Simple Layering

Use one-year-old suckers from the base of the parent plant. Notch or ring the underside of the branch to encourage root growth. Detach the new shoot and transplant it after a year of growth.

GARLIC
Allium sativum–Amaryllidaceae

Garlic is a hardy herbaceous perennial originally from southern Europe. The bulblets are used as seasoning in many cooked dishes or pickled foods. Garlic does not produce seed and may be propagated only by vegetative methods.

Garlic is hardy to all zones.

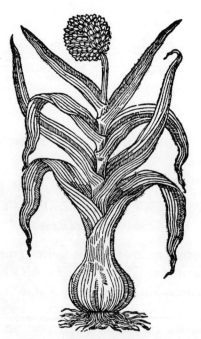

"GREAT MOUNTAIN GARLIC"
Allium sativum

Propagation
Bulblets
In the fall, the garlic plant produces a cluster of bulblets at its base. Dig the cluster, and cure it in the sun for a few days. Then overwinter it in a cool, dry spot, keeping moisture away to prevent mildew or rot. In the spring, break the cluster into individual cloves and plant them in the garden.

GOOD KING HENRY
Chenopodium bonus-henricus–Chenopodiaceae

Good King Henry is a herbaceous perennial from Europe. The entire young plant is used as a potherb. Dense clusters of tiny flowers on the erect flowering stalk are cross-pollinated by the wind.

Good King Henry grows well in all zones.

Propagation
Seed
Good King Henry will flower and fruit in the same manner as spinach each year. See seed propagation instructions for spinach.

GOOSEBERRY
Ribes spp.–Saxifragaceae

The gooseberry is a woody shrub armed with sharp spines and native to both America and Europe. Many varieties yield sharply sour, oblong or rounded berries that are exquisite in jams and jellies. The flowers are cross-pollinated by insects.

The gooseberry is hardy to zones 4–6.

Propagation
Hardwood Cuttings
See propagation instructions for the currant, the gooseberry's close cousin (Fig. 4).

Mound Layering
Again, see propagation instructions for the currant.

GRAPE
Vitis spp.–Vitaceae

The grape is a well-known perennial woody vine of both North America and the Old World that has been cultivated for centuries. The fruit is eaten fresh or may be fermented into wines of many types. Grapes are propagated entirely by vegetative methods. The culture of grapes is complex, and reference to Bailey and Bailey (1976) and Cox (1974a) should be made for additional details.

Grapes do best in zones 5–7 or in cooler parts of zone 8; some varieties have been specially developed for zone 9.

Propagation
Cuttings
In autumn, make 8–12″ cuttings at the tips of canes taken from well-ripened wood. Each cutting should contain 3 joints or nodes. Store the cuttings over the winter in moist sand. Plant them in the spring, right side up, making sure that the top bud is level with the ground. Cuttings will root in one season. Transplant them in the fall to a winter location that is protected from freezing. Transplant young plants to a permanent location the second spring.

Grafting
Whip-graft scions to rootstock grown from cuttings. In this manner it is possible to grow named varieties on hardy homegrown rootstock.

Layering
Simple or serpentine layering may be used to develop several new plants from lengthy canes. Separate the new plants after one year.

VINE
Vitis vinifera

GRAPEFRUIT
Citrus paradisi–Rutaceae

The nativity of grapefruit is unknown, but the large juicy fruit of the tree is familiar to almost everyone. Propagation depends upon vegetative means.

Grapefruit is hardy to zones 9–10.

Propagation

All members of the genus *Citrus* can be propagated the same way. See instructions for orange propagation.

GUAVA
Psidium spp.–Myrtaceae

The guava is native to the American tropics: strawberry guava *(Psidium cattleianum)* from Brazil; Costa Rican guava *(Psidium friedrichsthalianum)* from Central America; and the common guava *(Psidium guajava)* from tropical America. All are trees cultivated for their edible fruit, eaten fresh or used in jellies and preserves. The white flowers are cross-pollinated by insects.

The guava does well in zones 9 and 10 and prefers a tropical climate, although it is able to withstand a few degrees of frost.

Propagation
Seed

Sow seeds saved from tree-ripened fruit in flats or pans of light, sandy loam. Grow seedlings in pots for one season; then transplant them to permanent location.

Budding

Shield- or patch-bud choice varieties to home-grown seedling stock.

Cuttings
Stem cuttings may be rooted in moist sand.

HERB-PATIENCE
Rumex patientia–Polygonaceae

Herb-patience is a herbaceous perennial from Europe. The leaves are used as greens. The flowers are tiny and packed in dense clusters along the erect spikelike stem. Like many members of this family, the flowers are cross-pollinated by the wind.

Herb-patience is hardy to all zones.

Propagation
Seed
Elongated panicles of seed are produced each year in the late summer. The seeds are winged and are inherently dry when they appear on the plant. The mature stalks may be cut in the fall and dried further on a canvas in the sun. Remove the seeds by rubbing the panicles through your hands. Winnow out crushed wings and other debris. Dry one day and store.

HICKORY
Carya spp.–Juglandaceae

Hickory is a perennial, deciduous, woody tree native to North America. There are several species, all yielding edible nuts. The plants are monoecious, bearing unisexual catkins that are cross-pollinated by the wind.

Hickory does well in zones 5–8.

Propagation
Grafting
Graft scions from superior-bearing trees to rootstock seedlings of natives grown from seed.

Budding
Ring or annular budding is best for hickory. Use rootstock as in grafting.

Seed
Stratify the seed over the winter and plant it in the spring. You may choose to sow in autumn and let mother nature do the stratifying, provided you can protect the unsprouted seed from rodents. Transplant young seedling trees to permanent locations within two years; damage to the deep taproot may result if you move older trees.

HORSERADISH
Armoracia rusticana–Cruciferae

Horseradish is a hardy, deep-rooted perennial herb originally from southeastern Europe. The grated root is used as a pungent seasoning in dishes, salads, and sandwiches. The plant almost never flowers; hence seed is nonexistent.

Horseradish is hardy to zones 3–8.

Propagation
Root Division
Dig up the entire plant in the autumn. Cut the side roots, making the cuttings as long as possible. Make the top cut flat and the bottom cut slanted to help distinguish top from bottom. Tie the cuttings in bundles and pack them in sand for the winter. Store the cuttings in a cool, moist spot such as the root cellar, taking care to prevent withering. In the spring, plant root sections with the flat (thicker) end up, right at ground level. Each root piece will develop into a new plant (Fig. 5).

Crown Division
After overwintering the plant, cut the main root longitudinally into 4 pieces, leaving each with a part of the root and a part of the

HORSERADISH
Armoracia rusticana

crown. Plant each piece so that the top is at ground level. This method provides robust plants for the next fall harvest but does not allow you to use the current crop to make horseradish sauce, since the entire plant is divided into new plants.

JUJUBE
Zizyphus jujuba–Rhamnaceae

Jujube is a slow-growing, thorny, deciduous tree of New Zealand, Asia, and Africa that yields edible fruit.

The tree thrives in the hot climates of zones 9–10.

Propagation
Grafting
Whip-graft dormant scions on seedling stock raised from cleaned and stratified seed.

Seed
Seedlings are not quite true to type but may produce decently bearing trees in addition to rootstock (Riotte 1973a).

KALE OR BORECOLE
Brassica oleracea–Acephala Group–Cruciferae

Kale, a herbaceous perennial originally from Europe, was known to the ancient Greeks. Its leaves are used as greens. The flowers are cross-pollinated by insects, and the seeds remain viable for 4 years.

Kale is a cool-season crop thriving in zones 3–8.

Propagation
Seed
Kale is of the same species as cabbage, collards, and other varieties of *Brassica oleracea* and behaves similarly. See instructions for cabbage propagation, but take care to eliminate any crossing problems.

SEA KALE
Crambe maritima–Cruciferae

Sea kale is a hardy herbaceous perennial from coastal Europe grown for its young leaves and shoots, which are used as greens. The flowers are cross-pollinated by insects, and the seeds remain viable for one year.

Sea kale is hardy to zones 3–8 and grows primarily in coastal areas.

Propagation
Seed
Save the fruits (siliques) when they are thoroughly mature and dry. Store them in this condition. Plant the entire globular, single-

SEA KALE
Crambe maritima

seeded fruit in the spring. Carry the seedling in a temporary bed for one year, then transplant it to a permanent site.

Root Cuttings
Cuttings 4–5″ long should be made from vigorous roots and planted directly in the field, with the top of root at ground level. Do this in either spring, summer, or fall. A good crop may be expected the second year after planting.

KOHLRABI
Brassica oleracea–Gongylodes Group–Cruciferae

Kohlrabi is of unknown origin. The low biennial is grown for the tuberlike enlargement of the stem above the ground. The flowers are cross-pollinated by insects, and the seed remains viable for 3 to 5 years.

Kohlrabi is a hardy, cool season crop that does well in zones 3–8.

Propagation
Seed

Seedstalks are produced in the second year, as they are in many crucifers. Overwinter the first-year plant or stump in the garden, keeping it heavily mulched in cold climates. Or store the plant in the root cellar and plant it outdoors in the spring. Thereafter treat kohlrabi according to the instructions for cauliflower propagation.

KUMQUAT
Fortunella spp.–Rutaceae

The kumquat comes originally from southeastern China. The evergreen tree or shrub bears small citruslike fruits.

The kumquat is somewhat hardier than *Citrus* fruits and does well in zones 8–10.

Propagation
Grafting

Graft dormant scions by any standard method to orange or lemon seedling stock grown from stratified seeds or seeds planted fresh.

Budding

Use any budding method to bud kumquat to the same stock as above.

LEEK
Allium ampeloprasum–Porrum Group–Amaryllidaceae

The leek is a hardy, herbaceous biennial of Europe and western Asia and known to the ancient Greeks and Romans. It has been

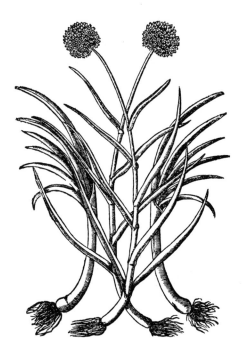

LEEK
Allium ampeloprasum

cultivated since ancient times for its soft bulb and leaves, which are eaten raw in salads, or steamed as a potherb. The flowers of the leek are cross-pollinated by insects and the seeds will remain viable for 2 to 3 years.

Leeks grow well in all zones.

Propagation
Seed
The leek, like the onion, produces a globular umbel of small flowers, which set seed. When the leek is in its second year, treat it according to the instructions for propagating the onion.

LEMON
Citrus limon-Rutaceae

The lemon is a tree that is probably native to Asia. It yields an oblong fruit with a sour flavor that is used in flavoring many foods and cold drinks.

The lemon thrives in zones 9–10.

Propagation

See instructions for orange propagation.

LENTIL
Lens culinaris-Leguminosae

The lentil is a herbaceous annual from southern Europe. Its seeds are used in soups and other hot dishes. The flowers are self-pollinated, and the seeds remain viable for 4 years. The lentil grows in zones 5–10.

Propagation
Seed

Allow the pods to mature and dry on the vine, or pull the entire mature plant before the first frost and hang it in a warm, dry room. Pick the pods when they are thoroughly dry, and store the seed in the pod, as this yields a better germination rate the following spring (Emery 1975). Seed may also be shelled from the pods and dried in pie dishes to be stored in envelopes or jars.

LETTUCE
Lactuca sativa-Compositae

The lettuce originated in Europe and Asia and has been cultivated for more than 2,500 years. It is a hardy herbaceous annual and comes in many varieties. Asparagus lettuce yields a long, thick main stem that is eaten. Cos or romaine lettuce and curled lettuce (all lettuces are considered the same species) bear loosely clustered leaves, while the leaves of head lettuce form a dense capitate rosette. All forms of lettuce are used in salads and sandwiches. Lettuce is mostly self-pollinated by insects, although cross-pollination can occur. Seed will remain viable for 5 to 6 years.

Lettuce thrives in all zones; it is a hardy cool-season crop and runs to seed in hot weather.

Propagation
Seed

All forms of lettuce can be treated similarly. Select a parent plant from the plants last to "bolt." For head lettuce, pull back the leaves to expose the growing tip when the head nears full size; otherwise the flowering stem may not be able to elongate for seed production. With all nonheading varieties, this treatment is not necessary. After flowering, nonheading lettuce produces many featherlike clusters of seeds with fuzzy attachments known as "pappus." Harvest the entire plant when half of the flower clusters are in this fuzzy "feather" stage. Dry the plant for 3 to 5 days in full sun or in paper bags. When the plant is dry, place the plant upside-down in the sack and shake it; the seeds will fall to the bottom. You can screen the seeds to remove any foreign matter; use a small-mesh screen since lettuce seeds are very tiny. However, home-grown lettuce seed need not be perfectly clean since the debris will not inhibit germination the next spring.

Because lettuce is mostly self-pollinated, you can grow more than one variety at a time for seed and achieve a high degree of fidelity. However, lettuce varieties cross readily with wild prickly lettuce *(Lactuca serriola)* and should be isolated from it if it is in the region.

LIME
Citrus aurantifolia–Rutaceae

The lime is a tree from India and southeastern Asia. The small, greenish-skinned fruit yields a sour juice that is used to flavor food and cold drinks.

The lime is hardy to zones 9–10.

Propagation

See instructions for orange propagation.

LOQUAT
Eriobotrya japonica–Rosaceae

The loquat is an evergreen tree native to China and cultivated for its small yellow, pearlike fruit which is eaten fresh or used in pastries and preserves. In the fall, the fragrant white flowers appear in terminal clusters and are cross-pollinated by insects. The fruit ripens by spring, and the seeds remain viable for 2 to 4 years.

The loquat grows well in frost-free, subtropical or tropical climates and is found in zones 9 and 10.

Propagation
Seed

Unimproved trees may be grown from seed taken from ripe fruits and germinated in flats of moist sand.

Budding

The best results are achieved by using any standard method to bud improved varieties on seedling stock.

MACADAMIA NUT
Macadamia integrifolia–Proteaceae

The Macadamia nut, or Queensland nut, is native to Queensland and New South Wales. The trees are cultivated for their edible nuts. The white flowers are probably insect-pollinated. Seed viability is unknown; germinate seed within one year.

The macadamia does well in tropical to subtropical temperatures, zones 9 and 10, but can withstand light frosts.

Propagation
Seed

Germinate seeds singly in pots of light soil, and grow them in a protected area for one season; then transplant the young trees to a permanent location.

MANGO
Mangifera indica–Anacardiaceae

The mango tree is native to tropical Asia and yields soft-textured fruits that are edible and sweet. The flowers are cross-pollinated by insects.

The mango thrives in tropical climates and is hardy to zone 10.

Propagation
Seed

Each fruit contains one very large seed, which can be removed from the pulp, cleaned, and planted in a flat or pot indoors. After one year, transplant the young tree to a permanent location outside. Mango trees may be grown only in areas without frost. The offspring from the seed may not be exactly true to type.

Budding and Grafting

Start seedling stock as above; then use any method of budding or grafting to join scions or buds from a named variety to the stock. Perform these operations during the colder parts of the year.

Hardwood Cuttings

Take cuttings from named varieties during the colder seasons; propagate these in pots the first year. Transplant the young plants to their permanent locations by the second season.

MANIOC OR TAPIOCA
Manihot esculenta–Euphorbiaceae

Manioc, known also as tapioca, yuca, or cassava, comes from eastern equatorial South America, especially Brazil. This perennial herbaceous shrub yields tuberous roots from which a flour or beads of tapioca can be made. The tubers can also be cooked and eaten like potatoes. Manioc flourishes in zones 9–10.

Propagation
Stem Cuttings

Take 4–10" of mature stem pieces from the plant, and plant them immediately 4' apart in each direction. In temperate climates, bury the entire cane beneath the ground to overwinter, and take cuttings in the spring.

MARTYNIA OR UNICORN PLANT
Proboscidea fragrans–Martyniaceae

Martynia is a low, spreading, herbaceous perennial native to the eastern United States. It is grown for its seed pods, which are a curiosity when mature and which may be pickled when young. The flowers are cross-pollinated by insects, and the seeds remain viable for 1 to 2 years.

Martynia is hardy to zones 5–10.

Propagation
Seed

Let chosen fruit mature on the plant. Then pick the fruit, separate the seeds out by hand, and clean it in water. Dry the seeds thoroughly and store them until you are ready to start the new plants in the spring.

MILLET
Panicum spp. and *Setaria spp.*–Gramineae

Millets are herbaceous annuals and perennials from the United States and East Indies. They are grown for their seeds, or grains, which are usually cooked whole or ground into flour. The small, inconspicuous flowers are cross-pollinated by the wind. Seed viability diminishes rapidly after one year.

Millet is hardy to all zones.

Propagation
Seed

Nearly all grains breed true to type, so simply let heads mature on the plant and pick in the fall before the heads break apart. Thresh the heads and winnow the grain to remove any debris. Thoroughly clean grain to be used as food; seed grain need not be as clean. Dry the seed a few days more and store.

MUSKMELON
Cucumis melo–Cucurbitaceae

The muskmelon and its relatives are all annual herbaceous vines bearing large, round, sweet and juicy melons. Muskmelon comes from Asia, as does the netted melon (Reticulatus Group), but the cassaba and honeydew (Inodorus Group) originated in Africa and southeast Asia. All of these are monoecious, and flowers are cross-pollinated by insects. Seed viability is 4 years.

Muskmelon is a crop hardy to zones 5–10.

Propagation
Seed

Let a chosen melon mature on the vine. A slight crack at the point of attachment to the stem indicates the maturity of both the fruit

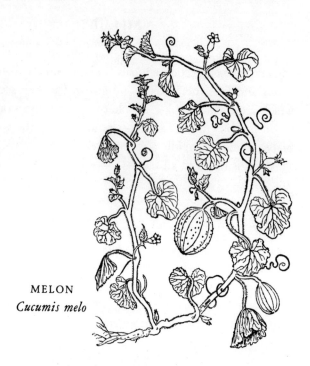

MELON
Cucumis melo

and its seeds. Pick the melon, cut it in half, and scrape out the seeds. Ferment the seed pulp in a crock or wooden vessel for 4 to 7 days at 70°F, stirring frequently. Add a little water as needed. The seeds will sink to the bottom, and the pulp will float. Decant the pulp and liquid; then dry the seeds on a thin screen in the sun. Store them for planting in the spring.

Be careful to prevent crossing with other varieties of *Cucumis melo* or *Cucumis sativus*.

LEAF MUSTARD
Brassica juncea–Cruciferae

Leaf mustard and its variety, southern curled mustard (var. *crispifolia*), come from the Old World. Both are herbaceous annuals whose leaves are used as greens. The flowers are cross-pollinated by insects, and the seeds remain viable for 4 to 5 years.

Leaf mustard is hardy to zones 3–8.

Propagation
Seed

The plant will bloom in the summer, and the seeds will mature in siliques on the elongated seed stalk by fall. Then propagate leaf mustard according to the instructions for any other member of the genus *Brassica*, such as head cabbage, cauliflower, or broccoli.

NECTARINE
Prunus persica var. *nucipersica*–Rosaceae

The nectarine is a tree from China that bears a smooth-skinned stone fruit very similar to the peach. The flowers are cross-pollinated by insects.

The tree is hardy to zones 5–8.

Propagation
Budding and Grafting

The nectarine is really a variety of peach with a smooth skin. It may be propagated in the same manner as its close cousin, and intergrafting with the peach can be readily accomplished.

OATS
Avena sativa–Gramineae

Oats probably originated in Tartary in west-central Asia. The plant is a herbaceous annual yielding a grain that is used as food both for human beings and for livestock. The tiny flowers are cross-pollinated by the wind.

Oats grow in zones 3–8.

Propagation
Seed

Oats may be propagated like any of the grains. See instructions for propagating barley.

OKRA OR GUMBO
Abelmoschus esculentis–Malvaceae

Okra is a herbaceous annual found in hot climates and originally from tropical Africa or Asia. The young fruit is cooked as a dinner vegetable. Okra blossoms are self-fertilized, and the seeds have a viability of 1 to 3 years.

Okra is found in zones 8–10.

Propagation
Seed

Allow the selected fruit to mature on the vine. Then pick it, cut it in half, and remove the seeds. Wash the seeds thoroughly to remove the mucilaginous material or, if necessary, ferment them in a crock for 2 days. Then rinse the seeds and dry them thoroughly on screens. Store them until spring.

OLIVE
Olea europaea–Oleaceae

The olive comes from the eastern hemisphere, Asia Minor, where it has been grown since prehistoric times. The tree bears bitter, black, single-pitted fruits that become palatable following special treatment with lye.

Olives grow well in zones 9–10 and are especially adaptable to Mediterranean climates.

Propagation

Softwood Cuttings

Take cuttings 4–5" long, retaining and cutting back only the leaves on the tip. Start the cuttings in sand in the nursery, or pot them in a south window. After the shoots take root, transplant the young trees to a permanent location.

Budding and Grafting

Attach a bud or scion to seedling stock by whip grafting or shield budding. Seedling stock may be started from fresh, untreated olive pits. Cutting off the pointed end of the pit will enhance seed germination.

ONION

Allium cepa–Amaryllidaceae

Onions originated in western Asia and have been cultivated from the earliest times. All forms of onion are hardy, cool-season, herbaceous biennials or perennials. The bulbs and leaves are used as food, mainly for seasoning. Some are cross-pollinated by insects; others are self-pollinated. Seed viability is 1 to 2 years.

Onions do well in all zones.

Propagation

Seed

Each onion plant (Cepa Group) elongates to produce a globular cluster of small flowers, which later matures, forming dry capsules of small black seeds. Harvest the capsules when they first begin to open, exposing the seeds. Cut the heads and dry them on canvas in the full sun for 3 weeks. The seeds will then shake out very easily, like salt from a salt shaker. Collect and store them until spring.

Sets

Sets may be produced in one season by planting seed thickly and maturing the crowded plants early. The small bulbs then go dormant and will resume growth when replanted—spaced farther apart—the following spring. Each set will produce a good bulb the second season. Dig sets in the fall; cure them for several days in the sun and store them over the winter in a cool, dry place.

Bottom Sets

Multiplier onions (Aggregatum Group) develop divided bulbs in the ground, which may be harvested in the fall, cured, stored in a cool, dry place over the winter, and divided and set out in the spring.

Top Sets

In the top or tree onion (Proliferum Group), bulblets appear in the flower cluster in the place of blossoms or seeds. This is an asexual process. Simply pick the cluster when it is full, break it apart, cure it in the sun for several days (Tirrell, 1969), and store in a cool, dry place.

WELSH ONION
Allium fistulosum-Amaryllidaceae

Also a herbaceous biennial from Asia, the Welsh onion yields bulbs and leaves used for food and as a seasoning. It behaves exactly like the common onion in terms of flowering and seed production.

The Welsh onion thrives in all zones.

Propagation
Seed

See instructions for onion seed propagation.

ORACHE

Atriplex hortensis–Chenopodiaceae

Orache is a herbaceous annual from Asia, cultivated for its greens, which are used as a potherb. The flowers are tiny and inconspicuous, densely clustered on the erect flowering stem. Cross-pollination occurs by the wind, and the seeds remain viable for 5 years. Orache is hardy to all zones.

Propagation
Seed
See instructions for propagating spinach, page 127. Remember orache flowers are perfect (both sexes are present), which means that a single plant may produce seed.

ORANGE

Citrus sinensis–Rutaceae

The orange tree, as well as the entire genus *Citrus*, comes from tropical and subtropical regions of Asia and the Malay Archipelago, or southeastern Asia. The trees are grown for their juicy fruit. Seed viability is high for one year.

Oranges thrive in zones 9–10.

Propagation
Shield Budding
Bud the desired variety on 2–3-year-old seedling stock of the same or related species (sour orange, trifoliolate orange, or rough lemon stocks are best). Budding should be done in autumn to allow a full growing season after growth begins. In the spring, just before the buds push out, cut the stock with a neat sloping cut. Stake the growing bud until it reaches 2–2½′ tall; then cut the top of the bud growth to force branching. Citrus trees grow shapely, sym-

ORANGE
Citrus sinensis

metrical heads needing comparatively little further pruning. Seedling stock may be grown from seeds taken directly from the orange and planted immediately in rich soil. Citrus fruits can also be integrafted.

Grafting methods, as well as layering and planting cuttings, can also be employed to propagate the orange.

PAPAYA
Carica papaya–Caricaceae

The papaya is a perennial, herbaceous tree of tropical climates, originating in tropical America. The fruits are soft and fleshy.

Some plants bear perfect flowers (both sexes present); others may be dioecious. Insects effect cross-pollination. Seed viability diminishes after one year.

The papaya requires tropical temperatures but can withstand a few degrees of frost. It is hardy to zone 10.

Propagation
Seed
Clean the seed from the fruit and plant it indoors in January. Set the young plant out in March. The plants will bear for two years, after which they should be replaced. If the plants are dioecious, both sexes will be necessary to produce fruit and seed.

Cuttings
Papaya may also be propagated by planting cuttings taken from the tips of branches (Brinhart 1969). Again, plant both sexes unless perfect plants are the source.

PARSLEY
Petroselinum crispum–Umbelliferae

Parsley and turnip-rooted parsley (var. *radicosum*) have been cultivated for over 2,000 years in Europe and western Asia, their region of origin. Both are herbaceous biennials furnishing leaves for flavoring and garnishing. The flowers are cross-pollinated by insects, and the seeds remain viable for 1 to 3 years.

Parsley grows in all zones.

Propagation
Seed
Compound umbels form during the second year. The plant must be overwintered by mulching (it is somewhat hardy). After the blooms appear the second summer, treat parsley according to the instructions for carrot propagation.

PARSNIP
Pastinaca sativa–Umbelliferae

The parsnip is native to Europe and was known to the ancient Greeks and Romans. It is cultivated for its edible taproot. The tiny flowers are cross-pollinated by insects, and the seeds remain viable for 1 to 2 years.

Parsnips grow in all zones.

Propagation
Seed

Parsnip is a biennial like the carrot and may be treated in the same manner for saving seed.

PEA
Pisum sativum–Leguminosae

Peas apparently originated in Europe and Asia, but their prototype has never been found. The ancients grew these herbaceous annuals. Garden peas (var. *sativum*) are grown for their young seeds, which make a fine dinner vegetable. The edible-podded pea (var. *macrocarpon*) yields fruit that is entirely edible, pod and all. All peas are self-pollinated, and the seeds will remain viable for 3 years.

Peas are a cool-season, frost-hardy crop and do well in all zones.

Propagation
Seed

All forms of the pea may be propagated the same way. Let peas mature in the pod; harvest them when the pods are dry and hard. Cut the vines at the base, and let them dry in the sun for one week. Then shell the peas from the pods and dry them further in the sun or in pie dishes indoors. The seeds may then be stored in a jar until spring.

COWPEA
Vigna unguiculata–Leguminosae

The cowpea, or southern pea, is a herbaceous annual from Asia or central Africa. The seeds are used as human food, and the foliage is consumed as livestock fodder. Cowpea flowers are self-pollinated, and the seeds remain viable for 3 years.

Cowpeas enjoy mild climates and are hardy to zones 7–10.

Propagation
Seed
The cowpea is really a bean. See instructions for kidney bean propagation.

PEACH
Prunus persica–Rosaceae

The peach dates back to 2000 B.C., when the Chinese cultivated the deciduous tree for its delicious stone-fruit. The fruit develops from flowers cross-pollinated by insects. Seed viability is one year.

Peaches do well in zones 5–8 and the cooler parts of zone 9.

Propagation
Shield Budding
It is best to attach a bud to seedlings that have never been transplanted. Grow seedling stock from seeds taken from peach pits. Plant the seeds outside in the fall, or stratify them over the winter to be planted the following spring. After 1 or 2 years, insert buds from a desired variety onto the seedling stock. Do this in late summer or early fall, after which the buds will remain dormant until spring. Then the stock above the bud should be cut to force bud growth. Remove all other sprouts from the stock. The tree will normally begin to bear in 3 to 6 years.

PEANUT
Arachis hypogaea–Leguminosae

The peanut is a herbaceous annual from Brazil, cultivated for its seeds. Peanuts flower above the ground, self-pollinate, and develop their fruit and seed underground. Seed viability is one year.

Peanuts prefer a long, mild season and do well in zones 7–10.

Propagation
Seed

Dig the entire peanut plant in the late fall before the first frost. Invert the plant and dry it thoroughly in the sun. Remove the fruit and store them whole. If the pods are very hard, remove them before planting; otherwise, plant the entire pod (Rozell 1970).

PEAR
Pyrus communis–Rosaceae

The pear tree is native to Europe and western Asia and was known to the Greeks and Romans as early as 300 B.C. The flowers are cross-pollinated by insects, and mature as the familiar sweet, juicy, pome fruits we know. Seed viability is one year.

Pears thrive in zones 5–8.

Propagation
Whip Grafting

Graft the desired variety on whole roots, making the union at the crown. Use seedling stock raised from pear seeds that have been cleaned, stratified, and germinated the year before.

Budding

Use the shield-budding method, best performed in the autumn. Seedling stock may be grown from seed as outlined above.

For dwarf varieties, graft or bud onto quince seedling or rootstock. The pear, apple, and quince may be intergrafted.

PECAN
Carya illinoinensis–Juglandaceae

The pecan is native to North America, in the region from Indiana to Mexico. The deciduous tree yields nuts similar to walnuts, but smaller and smoother-shelled. The flowers are cross-pollinated by insects, and the plants are often incompatible.

The pecan will grow in zones 6–9, and hardy varieties may be grown in zone 5.

Propagation
Budding and Grafting
Grow seedling stock from stratified pecan nuts. Bud named varieties or whip-graft scions to these seedlings after one year.

Many varieties of pecan show self-sterility, therefore in any plantation several varieties should be planted. Also, the plants have long taproots and should be transplanted carefully.

PEPPER
Capsicum annum–Solanaceae

Peppers are herbaceous perennials usually grown as annuals, especially in temperate climates. The bell pepper (Grossum Group) originated in South America, the hot pepper (Longum Group) in Peru, Central America, and South America. Both are grown for their fruit, which is used in cooking, salads, various pickling recipes, or simply eaten raw, either whole or sliced. The flowers are usually self-pollinated, although crossing by insects does occur. The seed remains viable for 4 years.

The pepper grows in all zones, although it requires a long growing season.

PEPPER
Capsicum annuum

Propagation
Seed

Since one fruit contains many seeds, a single specimen will be sufficient for the next year's crop. Cut the pepper when it is red, and scrape the seed from the inside. Wash and dry the seed immediately; then store it until early spring.

PERSIMMON
Diospyros spp.–Ebenaceae

The common persimmon tree *(Diospyros virginiana)* is native to the eastern United States, whereas the Japanese persimmon *(Diospyros kaki)* comes from the Orient. Both species bear soft, juicy fruit with a rather smooth texture but a "puckery" flavor when unripe. The plants are monoecious.

Plant *D. kaki* in zones 8–10, and *D. virginiana* in zones 5–10.

Propagation
Shield Budding
Cut long, heavy buds just before the rootstock becomes dormant in winter, and graft the buds onto native stock.

Whip Grafting
In winter, graft the desired scion on native stock just below the ground surface. Seedlings should be one or two years old before grafting.

Seeds, cuttings, and layering are also successful propagation methods for the persimmon.

PINEAPPLE
Ananas comosus–Bromeliaceae

The pineapple is a herbaceous perennial from tropical America. The fruit, which hardly ever contains seeds, results from the development of many individual flowers clustered in a spikelike mass.

Pineapple prefers tropical climates and is hardy to zone 10.

Propagation
Suckers
Shoots develop in the axils of leaves below the fruiting stem near the ground. Cut these shoots and root them immediately in potting soil.

Slips
Similar to green cuttings, these shoots develop just below the fruit along the stem. Cut these and root them as you would suckers.

Crown of Fruit
The leafy top of the fruit will root. Cut it from the fruit and place it in water, or keep it moist in sand. The new plant will come to maturity later than those grown by suckers or slips.

Crown Division
The stump of an old plant may be cut longitudinally, taking part of the old stem, crown, and root with each piece. Plant these pieces to form new plants.

PLUM
Prunus spp.–Rosaceae

Plum trees of many types have come from several parts of the world, including southwestern Asia, Japan, and North America. All are deciduous trees yielding juicy stone fruits. Many are cross-pollinated by insects; a few are self-pollinated.

Plums are hardy to zones 4–9, depending on the variety.

Propagation
Budding
Budding is the most successful method of propagating plums. Using *Myrobalan* stock *(Prunus cerasifera)* or homegrown peach stock (see instructions for peach), most budding techniques will succeed. Note that the plum and the peach may be successfully intergrafted.

POMEGRANATE
Punica granatum–Punicaceae

The pomegranate has been known from southern Europe and Asia since the earliest times. It is a perennial woody shrub or tree, and yields a globular, leathery-skinned fruit containing partitions full of juicy, red, pulp-covered, crystallike seeds.

The pomegranate is a tropical to subtropical fruit that does well in zones 9 and 10.

Propagation

Hardwood Cuttings

Take cuttings 1' long and ¼" in diameter, and root them in potting soil or sand. After one year, transplant the young trees to their permanent locations.

Layering

Use the simple layering method on branches that arch over and touch the ground or on branches that can be bent to the ground easily.

Seed

Seeds may be squeezed from the fruit and separated from the pulp in water. The seed will sink and the pulp will float. After decanting, stratify the seeds over the winter and plant them in the spring. The offspring will not be exactly true to type, but they will be good.

POTATO

Solanum tuberosum–Solanaceae

The potato originated in the Andes of Chile and Peru. This herbaceous annual, now cultivated throughout the world, yields the familiar tubers that are prepared in many ways. The flowers are self-pollinated, although potatoes are propagated mainly by vegetative means. The resulting seed is of importance only for developing new strains.

Potatoes thrive in all zones.

Propagation
Tubers

Used either whole or cut up into pieces (Fig. 19) with 1 to 3 eyes each, tubers are the primary means of potato propagation. Dig the tubers up in the fall, and store them over the winter in a cool, moist root cellar. By spring the eyes may have already begun to sprout and the potatoes wither, but they will still make good starts. Plant them at or slightly before the last frost of early spring. Use tubers that are free of fungus.

CUSHAW PUMPKIN
Cucurbita moschata–Cucurbitaceae

The cushaw pumpkin, or winter crookneck, is also of unknown nativity, perhaps from tropical America. It is an annual herbaceous creeping vine yielding edible fruit. Plants are monoecious and cross-pollinated by insects. The seeds remain viable for 4 to 6 years.

Like the pumpkin, the cushaw is a warm-season crop and is hardy to all zones.

Propagation
Seed

See instructions for field pumpkin propagation.

FIELD PUMPKIN
Cucurbita pepo var. *pepo*–Cucurbitaceae

The pumpkin is of unknown nativity, although it may have come from tropical America. It is an annual herbaceous creeping vine producing large pepo fruits that are used as vegetables. The plants are monoecious and are cross-pollinated by insects. Seed viability is 4 to 6 years.

The pumpkin is a warm-season crop hardy to all zones.

Propagation
Seed

Let the fruit mature until the frost has killed the foliage. Then pick the pumpkin, cut it in half, and scrape out the seeds and pulpy placenta. Place the whole mass in water in a wooden vessel or crock. Stir vigorously. The seeds will float and can be collected from the surface. Put the seeds in the full sun immediately to dry for several days, and store them.

Pumpkin will cross with other varieties of *Cucurbita pepo* and also with the cushaw pumpkin *(Cucurbita moschata)*, so you should separate these plants widely or grow them for seed in different years.

QUINCE
Cydonia oblonga-Rosaceae

The quince is native to Persia and Turkestan and has been cultivated since the beginning of recorded history. It is a deciduous tree yielding a pome fruit. Its flowers are cross-pollinated by insects.

Quinces grow in zones 5–8.

Propagation
Hardwood Cuttings

Make long cuttings in autumn, overwinter them in moist sand, and plant them to root in spring. The plants may be set out when they are a year old.

QUINCE
Cydonia oblonga

Budding and Grafting
Named varieties may be joined to seedling stock or rooted Angers cuttings.

Layering
In the absence of available rootstock, layer the flexible branches.

RADISH
Raphanus sativus–Cruciferae

The common radish and the Chinese radish (cv. *longipinnatus*) originated in Eurasia and China. Both are herbaceous annuals yielding a swollen, spicy taproot that is eaten raw, either whole or sliced in salads. The typical crosslike flowers that appear in summer are cross-pollinated by insects. Radish seeds remain viable for 5 years.

Radish is an early, cool-season crop that thrives in all zones.

Propagation
Seed

Choose a parent plant from those last to "bolt." You will then be selecting plants that will produce roots for a longer period of time each spring. Let the chosen plant flower and produce the siliques, which are somewhat inflated. Harvest the plants when the siliques are fully mature and drying. Cut the siliques from the stem, and dry them for 2 to 3 weeks in the full sun. Then shell the seeds by crushing the siliques between your hands. The seeds should fall readily from the dried crushed fruit. Sieve the seeds from the other debris, and dry them 2 weeks more before storing them for the winter.

Radishes cross with wild mustard and other varieties of radish, so separate your parent plants from possible crossing.

BLACK RASPBERRY, BLACKCAP
Rubus occidentalis-Rosaceae

The black raspberry, a perennial woody vine with biennial canes, comes from North America. The fruit forms in clusters of drupelets, familiar aggregates that are juicy and sweet. The flowers are cross-pollinated by insects.

Blackberries are hardy to zones 4–8.

Propagation
Tip Layering

Shoots are produced at the tips of the canes, which touch the ground and take root. This normal process may be encouraged by tip rooting (Fig. 10) or by covering the tips of the canes in early fall. Leave the new plant attached to the parent plant until the following spring, when you can cut and transplant it to a new location.

Simple Layering
In the spring, shoots can be bent to the ground and covered with soil. In two months new shoots will appear. Sever the shoot from the parent plant and move it to new location.

RED RASPBERRY
Rubus idaeus var. *strigosus*–Rosaceae

The red raspberry is a perennial woody vine with biennial canes that also comes from North America. The plants are cultivated for their fruit.

Red raspberries grow well in zones 3–8.

Propagation
Suckers
The raspberry freely propagates by growing new canes, which are suckers from the rootstock. Each of these canes is biennial, growing vegetatively the first year, flowering and fruiting the second year, and dying. Raspberry culture requires cutting the dead canes out and keeping the new canes from becoming so thick that the entire patch becomes a bramble. In the spring, when the first-year canes are coming up, detach them from the parent rootstock, taking some roots with them, and transplant them immediately to a new location. You may also transplant the first-year canes in the fall, after the plants become dormant. Raspberry patches may be started or enlarged in this manner.

RHUBARB

Rheum rhabarbarum–Polygonaceae

Rhubarb is a hardy herbaceous perennial that grows from a woody rootstock. It comes from the cooler portions of Asia, probably Siberia or Mongolia, and is grown for the petioles of the leaves, which are cooked to a sauce or used in jams. The elongating stalks produce dense clusters of small flowers that are cross-pollinated by the wind.

Rhubarb is hardy to all zones.

Propagation
Crown Division
Dig the entire plant in the fall or early spring and cut it into sections, including a part of the root with some perennating buds attached to each section. Plant each section immediately or, if dug in the fall, after overwintering in the root cellar. Divide the plants any time other than the season of active growth. Crown division might be done every 5 to 10 years to maintain a healthy and productive patch, in addition to supplying starts for a new patch.

Rhubarb does not come true from seed (Carlsen 1970).

RICE

Oryza sativa-Gramineae

Rice originated in the East Indies and has been cultivated for 4,000 years. It is an annual marsh herb yielding a grain that is used as human food throughout the world. The small flowers are cross-pollinated by the wind. The seeds remain viable for one year.

Rice requires aquatic conditions and grows well in zones 9–10.

Propagation
Seed
Collect seed from the mature spikes before they disintegrate. Then thresh and winnow. Store the dry seeds in jars. In the spring,

sprout the seeds in water and transplant them to an aquatic environment.

RUTABAGA
Brassica napus-Napobrassica Group–Cruciferae

Rutabaga is a hardy herbaceous biennial from Great Britain and northern Europe, cultivated for its edible root. The flowers are cross-pollinated by insects, and the seeds remain viable for 4 years.

Rutabaga can withstand hard frosts and requires a long season. It is hardy to zones 3–8.

Propagation
Seed

Grow the plant one season, then overwinter in the garden, heavily mulched in severe climates, or in the root cellar. Transplant the young to the garden in spring. The plant will flower and fruit during its second year. Treat rutabaga in the same manner as cabbage.

RYE
Secale cereale-Gramineae

Rye is native to Eurasia and has been cultivated since Roman times. The herbaceous annual yields seeds that are ground into flour or used for other purposes. The tiny, inconspicuous flowers in the spike are cross-pollinated by the wind. The seeds remain viable for one year.

Rye is hardy to zones 3–8.

Propagation
Seed

Treat rye as you would any other grain such as oats, barley, wheat, or rice.

SALSIFY, OYSTER PLANT, OR VEGETABLE OYSTER
Tragopogon porrifolius-Compositae

The salsify originated in southern Europe. It is a hardy herbaceous biennial grown for its edible root. The flowers are cross-pollinated by insects, and the seeds remain viable for 1 to 2 years. The salsify grows in zones 3-9.

Propagation
Seed
Seeds are produced in flowering heads during the second year. Treat salsify in the same way as you do chicory.

BLACK SALSIFY
Scorzonera hispanica-Compositae

The black salsify is a hardy herbaceous perennial from Europe, grown for its edible root and leaves. The composite flowers are cross-pollinated by insects in summer. Seed viability is 1 to 2 years.

The black salsify is somewhat hardy, living in zones 3-9, but the roots must be protected from freezing.

Propagation
Seed
Once established, black salsify will flower annually. When seed is desired, watch for heads and allow them to develop. To get seed from heads, treat them the same as chicory.

SPANISH OYSTER PLANT (SALSIFY), GOLDEN THISTLE
(*Scolymus hispanicus*-Compositae)

The Spanish salsify (better known as Spanish oyster plant) is a hardy herbaceous biennial from southern Europe and is grown for its edible taproot. The flowers are cross-pollinated by insects, and the seeds will remain viable for 1 to 2 years. Spanish oyster plant thrives in zones 3–9.

Propagation
Seed
Most members of the sunflower family behave similarly; treat Spanish oyster plant as you would chicory.

SAPOTA
Achras zapota-Sapotaceae

Sapota, or sapodilla, is an evergreen tree native to tropical America and grown for its delicious fruit. The solitary white flowers are cross-pollinated by insects, and the large, yellow-brown fruits contain black shiny seeds, which remain viable for 1 to 3 years.

Sapota enjoys a warm tropical climate and grows in zone 10.

Propagation
Seed
Germinate the seeds from tree-ripened fruit in individual pots. Transplant the young trees to a permanent location after one season.

Budding
Shield-bud selected strains on common seedling stock grown as indicated above.

SHALLOT
Allium cepa-Aggregatum Group–Amaryllidaceae

The shallot is a herbaceous perennial from western Asia. The hardy plant yields an edible bulb used for seasoning or eaten raw. The plant seldom flowers.

The shallot does well in all zones.

Propagation
Bulbs

Plant a bulb in the spring. By late summer the plant will have produced many bulbs from the first one. Dig up the cluster in late summer, and let it lie on the soil to cure in the sun for one week. Then separate the bulbs and store them in a cool, dry place over the winter. Repeat this cycle the following spring.

SORGHUM
Sorgum bicolor-Gramineae

Sorghum grains come from Africa or India. The herbaceous annual grasses yield seeds that are used as grain for human food. The stalks of one species of sorghum (*Sorghum bicolor*–Saccharatum Group) are used in the production of sugar. All varieties are cross-pollinated by the wind. Seed viability is 2 years.

Sorghum grows well in zones 6–10.

Propagation
Seed

Cut the mature heads in the late summer. Thresh and winnow the heads, and dry the seed thoroughly before storing. Store sorghum according to the instructions for any of the grains.

GARDEN SORREL
Rumex acetosa-Polygonaceae

Garden sorrel is a herbaceous perennial from Europe and Asia. The leaves are used as greens. Clusters of small flowers on an elongate stalk are cross-pollinated by the wind. The seeds remain viable for 2 years.

Garden sorrel does well in all zones.

Propagation
Seed

Garden sorrel is a close relative of herb-patience and may be treated in the same manner.

SOYBEAN
Glycine max-Leguminosae

The soybean is a herbaceous annual from China, India, and Japan and has been cultivated since ancient times. It is grown for its nutritious seeds. The flowers are self-pollinated, and the seeds will remain viable for 2 years.

Soybeans grow in zones 7–10.

Propagation
Seed

Treat this legume in the same manner as you would peas or beans.

NEW ZEALAND SPINACH
Tetragonia tetragonioides-Tetragoniaceae

New Zealand spinach is a succulent herbaceous annual originally from Japan, New Zealand, Australia, and South America. The leaves are used as a potherb and the flowers are cross-pollinated by the wind. The seeds will remain viable for 5 years.

New Zealand spinach does well in hot weather and is hardy to zones 4–10.

Propagation
Seed

The flowering stem of the New Zealand spinach continues blooming until the first frost. Harvest the entire plant when most of the seeds are mature—before the first frost. Dry the plant on canvas in the full sun for 2 to 3 weeks. The fruit will split; then you can simply shake the seeds out on the canvas, and collect and store them until spring.

SPINACH
Spinacea oleracea-Chenopodiaceae

Spinach is an annual herb originally from Asia and, most likely, Persia. The leaves are used as a potherb. The small flowers are cross-pollinated by the wind, and the seeds remain viable for 5 years.

Spinach is hardy to all zones; it is a cool-season crop, bolting quickly to seed when temperatures are high and days are long.

Propagation
Seed

Spinach is unique in that it produces four types of plant: extreme males, bearing only male flowers and usually the first to bolt; vegetative males, which produce no flowers and hence no pollen; monoecious plants bearing both female and male flowers separately; and female plants bearing only female flowers and, therefore, seeds. Save the seed from the plants that mature last because these are likely to be either female or monoecious. Treat spinach according to the instructions for beets or chard, except that it is an annual. When the plants begin to turn yellow after flowering and fruiting, harvest the entire plant and dry it in the full sun for 1 to 2 weeks before threshing. Dry the threshed seed 1 week further and then store it until spring. Both the prickly-seeded spinach (*S.o.* var. *oleracea*) and the round-seeded spinach (*S.o.* var. *inermis*) may be propagated this way.

In the absence of monoecious plants, you should grow some male plants to ensure pollination.

SUMMER SQUASH
Cucurbita pepo var. *melopepo*-Cucurbitaceae

Summer squash originated in tropical America and is now found in many varieties. All of these are herbaceous annuals yielding immature fruits that are either eaten raw or cooked as a vegetable. Squash plants are monoecious, and the blossoms are cross-pollinated by insects. The seeds collected will remain viable for 4 years. Squash is a warm-season crop hardy to all zones.

Propagation
Seed

Choose the fruit you want to use for seed and allow it to mature on the vine. When the frost has killed the foliage, pick the fruit and propagate it according to the instructions for propagating the field pumpkin. Remember that summer squash will cross with other varieties of *Cucurbita pepo* and with *Cucurbita moschata*.

WINTER SQUASH
Cucurbita maxima-Cucurbitaceae

The nativity of the winter squash is not known, but it may come from North America. It is a herbaceous annual vine that is grown for its mature fruit, which can be cooked and eaten in a variety of ways. As with other cucurbits, the plants are monoecious, and the flowers are cross-pollinated by insects. The seeds will remain viable for 4 years. Winter squash is a warm-season crop hardy to all zones.

Propagation
Seed

Treat the winter squash like the field pumpkin. The seed may be easily grown, since the winter squash will not cross with any other members of the gourd family. Varieties of the winter squash will cross with each other, however, so grow only one type each year for seed.

STRAWBERRY
Fragaria spp.-Rosaceae

The strawberry is a perennial stoloniferous herb whose varieties come from plants domesticated in several parts of the world, mainly the temperate regions of the northern hemisphere. All varieties of strawberry are cultivated for their fruit, which is eaten raw in desserts or used in pies, jams, or jellies. The leaves may also be brewed for tea. The flowers appear in early spring and are cross-pollinated by insects.

Strawberries do well in zones 4-8.

Propagation
Stolons

The parent plant produces abundant runners during the growing season. The runners root freely (Fig. 7), forming new plantlets with leaves as well as roots. Remove these new plantlets to prevent congestion in the strawberry bed and to provide plants for new beds. Cut the plantlet from its parent, and plant it so that its crown is just at ground level. This can be done in late fall in warmer climates or in early spring in cooler climates (Berlin 1969).

SUNFLOWER
Helianthus annuus-Compositae

The sunflower is a robust herbaceous annual from North America and is cultivated for its large seeds. The flowers are cross-pollinated by insects, and the seeds will remain viable 3 to 4 years.

Sunflowers are hardy to all zones, during the frost-free season.

Propagation
Seed

The large head of the sunflower produces hundreds of seeds, which you can eat or use to grow the next crop. Heads should be cut when they are fully mature. Dry them fully on canvas in the sun. When the heads are dry, rub the seeds from the heads and dry them further for one week on screens. Then store until spring.

SWEET POTATO
Ipomoea batatas-Convolvulaceae

The sweet potato is a trailing herbaceous perennial found in regions with long growing seasons. It comes from tropical America and the West Indies. The plants are cultivated for their edible tubers. The flowers are cross-pollinated by insects.

The sweet potato thrives in zones 7–10.

Propagation
Sprouts

In early spring, cut a tuber in half lengthwise, plant in a hotbed, and cover it with 2" of light soil. Sprouts will grow from the tuber in abundance and should be removed when they are 8–10" long. Each sprout will bear its own roots. Set the sprouts out in the desired location after all danger of frost is past. To time the setting of plants closely with the last frost, start the tubers about six weeks ahead of the anticipated planting date. Tubers may be overwintered in baskets or boxes in a warm, well-ventilated area with a temperature no less than 50°F. Preparation for overwintering consists merely of digging up the tubers from the current crop and drying them in the full sun for a couple of hours before storing.

Although this method is asexual, great variation in the offspring may result. This is unusual and results in variable tubers in the next crop.

Cuttings

Take vine cuttings from the existing vine and root them immediately in a new location in the garden. Such cuttings, when in production, bring more uniform tubers (Tripp 1970). This method works nicely in the southern United States, zones 8–10, where a long growing season prevails.

TARO or DASHEEN
Colocasia esculenta-Araceae

Taro is a herbaceous perennial found in warmer climates and native to tropical Asia. It has been cultivated for over 2,000 years in the Orient for its globose tubers, which are cooked as a vegetable. The shoots of the plant can be served as blanched greens. There is no seed production since the flowering parts are sterile.

Taro prefers a warm, subtropical climate and flourishes in zones 9–10.

Propagation
Tubers

Dig up the tubers of an existing plant. Divide and replant these at any time other than during the period of rapid growth.

CURRANT TOMATO
Lycopersicon pimpinellifolium-Solanaceae

The currant tomato is native of Peru. It is a tender herbaceous perennial treated as an annual. It yields small fruits similar to those of the tomato. As with the common tomato, the flowers are self-pollinated.

The currant tomato is a warm-season crop hardy to zones 4–10.

Propagation
Seed
See instructions for the common tomato.

HUSK TOMATO
Physalis pruinosa-Solanaceae

The husk tomato, ground cherry, or cape gooseberrry is a stout herbaceous annual originally from the eastern United States. The fruit is used in pickles and preserves. The flowers are self-pollinated, and the seeds remain viable for 8 years.

Husk tomatoes are hardy to zones 5-10.

Propagation
Seed
The fruit of the husk tomato, a berry surrounded by a swollen papery calyx, resembles a Japanese lantern. Let the fruit mature, and peel off the calyx from each one. Then propagate the yellowish berries according to the instructions for the common tomato.

TOMATO
Lycopersicon lycopersicum-Solanaceae

The tomato, a tender herbaceous perennial that is usually grown as an annual in temperate climates, comes originally from South America. The common tomato is familiar, used raw in salads or cooked in various ways as a component of many hot dishes. The cherry tomato (var. *cerasiforme*), pear tomato (var. *pyriforme*), and potato-leaved tomato (var. *grandifolium*) are all varieties of the same species. Since all tomato plants are self-pollinated, varieties tend not to cross, although crossing is possible. The seed produced will remain viable for 4 years.

Tomatoes are hardy to zones 4-10.

Propagation
Seed

Tomatoes of all varieties readily volunteer and are easily propagated. Choose the fruit to be the "parent" and allow it to ripen thoroughly before picking. Gather these tomatoes, cut them in half crosswise, and look for thick, meaty, well-colored walls and a small seed cavity. When the desirable fruit is found, squeeze the seed into an earthen or wooden container (never use metal for any of the fermentation processes) and save the remaining pulp for canning or eating. You may also place whole tomatoes in the container and crush them gently without injuring the seeds. Add some water and let the mixture ferment for 24 hours until the mucilaginous and sticky covering comes off the seeds. Add more water and stir. The good seed should settle to the bottom, and the pulp, skins, and stems will float. Decant this mixture several times, then spread the clean seeds on paper, cloth, or other absorbent material to dry in the shade with open air circulation. When the seeds are fully dry, store them until spring.

Green Cuttings

Cut the tip portion of the stem of a tomato plant, taking about 1' of stem with 2 or 3 leaves. Root the cutting in potting soil.

TURNIP
Brassica rapa-Rapifera Group–Cruciferae

The nativity of the turnip is unknown—some authorities believe it is from Russia or Siberia, and others believe it is from Great Britain. It is a herbaceous plant, grown either as a spring annual or a summer biennial. The swollen root is eaten raw or cooked, and the leaves are cooked as greens. The flowers are cross-pollinated by insects, and the seeds remain viable for 5 years.

The turnip does well in zones 3–8 and is a hardy, cool-season crop.

Propagation
Seed

Sow the seed early in the spring to get a seed crop the same year. Seed sown later will not produce flowering plants until the second season. Turnip flowers and fruit are very similar to other members of the Cruciferae and may be handled in the same manner. Harvest the plants when the siliques are yellow but not dry; cut off the seed stalks and put them in paper bags in a warm, dry spot. The siliques will open after 2 to 3 weeks, and the seeds will fall out easily. After screening, dry the seeds further for a few days and then store.

BLACK WALNUT
Juglans nigra-Juglandaceae

The black walnut is native to eastern North America. This deciduous tree is cultivated for its hard-shelled nut. The plants are monoecious, and the catkins are cross-pollinated by the wind.

Black walnuts grow in zones 5–9.

Propagation
Grafting

Desirable scions may be grafted to seedling stock grown from native black walnuts (Moffett 1977). Whip-graft the scion to year-old seedlings at the crown. One year after planting the graft, set the tree in the orchard. At that time, cut it back to force head development. Any method of budding may also be used; this is best done in the fall.

ENGLISH WALNUT
Juglans regia-Juglandaceae

The English, or Persian, walnut is a native of Persia. It is cultivated for its large, meaty nuts. The plants are monoecious, and catkins are cross-pollinated by the wind.

The English walnut grows well in zones 5–9.

Propagation

Follow instructions for propagating the black walnut. For best results, graft the English walnut to black walnut rootstock.

WATERMELON
Citrullus lanatus-Cucurbitaceae

The watermelon originated in tropical and southern Africa. The long-running, herbaceous annual vine yields large, juicy fruit. The plants are monoecious, and the blossoms are cross-pollinated by insects. Seed viability is 5 years.

Watermelon grows from the warmer parts of zone 4 through zone 10 and requires a long, hot growing season.

Propagation
Seed

Select the parent plant on the basis of fruit color, texture, and flavor. Allow one fruit to mature until it has a yellowish color. Then harvest it and cut it in half. Separate the seeds from the central portion only—these seeds are the most robust and fully developed, being the oldest. Place the seeds and pulp on a screen to drain. Wash the seeds free of pulp, and dry them immediately on screens for a couple of weeks.

Watermelon will not cross with any other member of the gourd family except the citron melon (*Citrullus lanatus* var. *citroides*). Therefore you can grow watermelon for seed near most other cucurbit plants.

WHEAT
Triticum aestivum-Gramineae

Wheat is an ancient plant, probably originating in the Tigris-Euphrates Valley. The herbaceous annual is grown for its grain, which is ground into flour to make bread. The small, grasslike flowers are cross-pollinated by the wind.

Wheat is hardy to zones 3–7.

Propagation
Seed

The grain may be threshed and winnowed, then follow instructions for propagating any grain. Seed grain need not be as thoroughly cleaned as food grain before storing, but both must be dried. The seed of emmer, einkorn, Polish wheat, and spelt can be saved as it is for wheat.

YAM
Dioscorea alata-Dioscoreaceae

The yam is native to India and Malaya. It is a perennial, twining, herbaceous vine that thrives in warm regions. The plants are grown for their edible tubers. The flowers of this dioecious plant are small, and borne in spikes, and probably insect-pollinated.

The yam thrives in zones 9–10, preferring tropical to subtropical climates.

Propagation
Tubers

Use either whole, small tubers or tubers cut into pieces. Dry the cut pieces briefly in the sun to let a crust form on the fresh surface, and plant them at the beginning of the growing season. Yam tubers are stored according to the instructions for the sweet potato.

Crown Division

Cut the crowns of the large tuberous roots, taking a bit of root and stem with each piece. Plant the pieces immediately in the desired location to get new plants.

Botanical Classification of Common Food Plants

The food plants discussed in Part Two are listed in this table by family group, using the scientific names (following Bailey and Bailey 1976) in alphabetical order:

FAMILY (Common Name)
    Genus and *species* (Common Name)
        Variety, Group, or Cultivar (Common Name)

AMARYLLIDACEAE (Amaryllis Family)
    Allium ampeloprasum (Wild Leek)
        Porrum Group (Leek)
    Allium cepa (Onion)
        Aggregatum Group (Multiplier Onion, Potato Onion, Shallot)
        Cepa Group (Onion)
        Proliferum Group (Tree Onion, Top Onion)
    Allium fistulosum (Welsh or Spanish Onion)
    Allium sativum (Garlic)
    Allium schoenoprasum (Chive)

ANACARDIACEAE (Cashew Family)
Mangifera indica (Mango)

ARACEAE (Arum Family)
Colocasia esculenta (Taro, Eddo, Dasheen)

BETULACEAE (Birch Family)
Corylus spp. (Filbert or Hazelnut)

BROMELIACEAE (Bromelia or Pineapple Family)
Ananas comosus (Pineapple)

CARICACEAE (Papaya Family)
Carica papaya (Papaya or Pawpaw)

CHENOPODIACEAE (Goosefoot Family)
Atriplex hortensis (Orache)
Beta vulgaris (Beet)
  Cicla Group (Swiss Chard)
  Crassa Group (Garden Beet, Sugar Beet, Mangel)
Chenopodium album (Lamb's-Quarters)
Chenopodium bonus-henricus (Good King Henry)
Chenopodium quinoa (Quinoa)
Spinacea oleracea (Spinach)
  var. *inermis* (Round-Seeded Spinach)
  var. *oleracea* (Prickly-Seeded Spinach)

COMPOSITAE (Composite or Sunflower Family)
Cichorium endivia (Endive)
Cichorium intybus (Chicory or Succory)
Cynara cardunculus (Cardoon)
Cynara scolymus (Globe Artichoke)
Helianthus annuus (Common Sunflower, Mirasol)
Helianthus tuberosus (Jerusalem Artichoke, Girasole)
Lactuca sativa (Lettuce, all forms)
Scolymus hispanicus (Spanish Oyster Plant, Golden Thistle)
Scorzonera hispanica (Black Salsify)

Taraxacum officinale (Common Dandelion)
Tragopogon porrifolius (Salsify, Vegetable Oyster)

CONVOLVULACEAE (Morning Glory Family)
Ipomoea batatas (Sweet Potato)

CRUCIFERAE (Mustard Family)
Armoracia rusticana (Horseradish)
Barbarea vulgaris (Winter Cress)
Brassica juncea (Leaf Mustard, Mustard Greens)
   var. *crispifolia* (Southern Curled Mustard, Ostrich Plume)
Brassica napus (Rape, Colza)
   Napobrassica Group (Rutabaga, Swede)
Brassica oleracea (Cabbage)
   Acephala Group (Kale, Tree Kale, Collards, Cow Cabbage)
   Botrytis Group (Cauliflower, Broccoli)
   Capitata Group (Head Cabbage, Savoy Cabbage)
   Gemmifera Group (Brussels Sprouts)
   Gongylodes Group (Kohlrabi)
   Italica Group (Sprouting Broccoli)
   Tronchuda Group (Portuguese Kale or Cabbage)
Brassica rapa (Field Mustard)
   Chinensis Group (Pak choi, Pe-Tsai, Celery Cabbage)
   Rapifera Group (Turnip)
Crambe maritima (Sea Kale)
Lepidium sativum (Garden Cress)
Nasturtium officinale (Watercress)
Raphanus sativus (Radish)
   cv. *longipinnatus* (Chinese Radish)

CUCURBITACEAE (Gourd Family)
Citrullus lanatus (Watermelon)
Cucumis anguria (Bur Gherkin)
Cucumis melo (Melon, Muskmelon)
   Cantalupensis Group (Cantaloupe)

    Chito Group (Lemon Cucumber, Mango Melon)
    Inodorus Group (Cassaba, Winter Melon, Honeydew)
    Reticulatus Group (Netted Melon)
  Cucumis sativa (Cucumber)
  Cucurbita maxima (Autumn and Winter Squash and Pumpkin)
  Cucurbita moschata (Canada, Cheese, and Winter Crookneck Squash and "Pumpkins")
  Cucurbita pepo (Summer and Autumn Pumpkin and Squash, Gourd and Marrow)
    var. *pepo* (Field Pumpkin, Vegetable Marrow, Acorn Squash)
    var. *melopepo* (Bush Pumpkin, Bush Squash, including Cocozelle, Summer Crookneck, Pattypan, and Zucchini)
  Sechium edule (Chayote)

DIOSCOREACEAE (Yam Family)
  Dioscorea alata and other species (Yam)

EBENACEAE (Ebony Family)
  Diospyros spp. (Persimmon)

ERICACEAE (Heath Family)
  Vaccinium macrocarpon (Cranberry)
  Vaccinium spp. (Blueberry)

EUPHORBIACEAE (Spurge Family)
  Manihot esculenta (Manioc, Cassava, Tapioca)

FAGACEAE (Beech Family)
  Castanea dentata (American Chestnut)

GRAMINEAE (Grass Family)
  Avena sativa (Oats)
  Hordeum vulgare (Barley)
  Oryza sativa (Rice)
  Panicum spp. (Millet)

Secale cereale (Rye)
Setaria spp. (Millet)
Sorgum bicolor (Sorghum, Kafir Corn, Curra, Shallu, Sugar Sorghum)
Triticum aestivum (Wheat)
Zea mays (Corn)
    var. *praecox* (Popcorn)
    var. *rugosa* (Sweet Corn)

JUGLANDACEAE (Walnut Family)
Carya illinoinensis (Pecan)
Carya spp. (Hickory Nut)
Juglans cinerea (Butternut)
Juglans nigra (Black Walnut)
Juglans regia (Persian or English Walnut)

LAURACEAE (Laurel Family)
Persea americana (Avocado)

LEGUMINOSAE (Pea Family)
Arachis hypogaea (Peanut)
Glycine max (Soybean)
Lens culinaris (Lentil)
Phaseolus acutifolius (Tepary Bean)
Phaseolus coccineus (Scarlet Runner, Multiflora)
Phaseolus limensis (Lima Bean)
Phaseolus vulgaris (Kidney Bean, Bush or Pole Snap)
Pisum sativum (Pea)
    var. *macrocarpon* (Edible Podded Pea)
    var. *sativum* (Garden Pea)
Vicia faba (Broad Bean, Horse Bean)
Vigna aconitifolia (Moth Bean)
Vigna angularis (Adzuki Bean)
Vigna mungo (Urd Bean, Black Gram)
Vigna radiata (Mung Bean)
Vigna unguiculata (Cowpea, Blackeyed Pea)

LILIACEAE (Lily Family)
Asparagus officinalis (Asparagus)

MALVACEAE (Mallow Family)
Abelmoschus esculentis (Okra, Gumbo, Lady's Finger)

MARTYNIACEAE (Martynia Family)
Proboscidea fragrans (Martynia, Unicorn Plant)

MORACEAE (Mulberry Family)
Ficus carica (Fig)

MUSACEAE (Banana Family)
Musa paradisiaca (Edible Banana)

MYRTACEAE (Myrtle Family)
Psidium spp. (Guava)

OLEACEAE (Olive Family)
Olea europaea (Olive)

PALMAE (Palm Family)
Cocos nucifera (Coconut)
Phoenix dactylifera (Date)

POLYGONACEAE (Buckwheat Family)
Fagopyrum esculentum (Buckwheat)
Rheum rhabarbarum (Rhubarb, Pie Plant)
Rumex acetosa (Garden Sorrel)
Rumex patientia (Herb-Patience, Spinach Dock)

PROTEACEAE (Protea Family)
Macadamia integrifolia (Macademia Nut, Queensland Nut, Australian Nut)

PUNICACEAE (Pomegranate Family)
Punica granatum (Pomegranate)

RHAMNACEAE (Buckthorn Family)
Zizyphus jujuba (Jujube)

ROSACEAE (Rose Family)
   Cydonia oblonga (Quince)
   Eriobotrya japonica (Loquat)
   Fragaria spp. (Strawberry)
   Malus spp. (Apple and Crab Apple)
   Prunus armeniaca (Apricot)
   Prunus dulcis (Almond)
   Prunus persica (Peach)
      var. *nucipersica* (Nectarine)
   Prunus avium (Sweet Cherry)
   Prunus cerasus (Sour Cherry)
   Prunus spp. (Plum)
   Pyrus communis (Pear)
   Rubus idaeus
      var. *strigosus* (Red Raspberry)
   Rubus occidentalis (Black Raspberry, Blackcap)
   Rubus spp. (Blackberry)
   Rubus ursinus (Pacific Dewberry, Loganberry, and derivatives)

RUTACEAE (Rue Family)
   Citrus aurantifolia (Lime)
   Citrus limon (Lemon)
   Citrus paradisi (Grapefruit)
   Citrus sinensis (Orange)
   Fortunella spp. (Kumquat)

SAPOTACEAE (Sapote Family)
   Achras zapota (Sapota or Sapodilla)

SAXIFRAGACEAE (Saxifrage Family)
   Ribes nigrum (Black Currant)
   Ribes sativum (Red Currant)

SOLANACEAE (Nightshade Family)
   Capsicum annuum (Pepper)

Cerasiforme Group (Cherry Pepper)
Grossum Group (Bell Pepper)
Longum Group (Hot Peppers, including Long Red, Chili, and Cayenne)
Lycopersicon lycopersicum (Common Tomato, including many forms)
  var. *cerasiforme* (Cherry Tomato)
  var. *grandifolium* (Potato-Leaved Tomato)
  var. *pyriforme* (Pear Tomato)
Lycopersicon pimpinellifolium (Currant Tomato)
Physalis pruinosa (Husk Tomato, Strawberry Tomato, Dwarf Cape Gooseberry)
Solanum melongena (Eggplant)
Solanum tuberosum (Potato)

TETRAGONIACEAE (New Zealand Spinach Family)
Tetragonia tetragonioides (New Zealand Spinach)

UMBELLIFERAE (Parsley or Carrot Family)
Anthriscus cerefolium (Salad Chervil)
Apium graveolens
  var. *dulce* (Celery)
  var. *rapaceum* (Celeriac)
Daucus carota
  var. *sativus* (Carrot, in many forms)
Pastinaca sativa (Parsnip)
Petroselinum crispum (Parsley)
  var. *tuberosum* (Turnip-Rooted Parsley)

VITACEAE (Grape or Vine Family)
Vitis spp. (Vine Grape, in many forms)

Note: Crossing is possible between different species of *Brassica* (Cruciferae) and between varieties of any species therein. Normally one does not need to worry about crossing between any members of Solanaceae; however, due to renegade bumblebees, groups within *Capiscum annuum* and varieties of *Lycopersicon Lycopersicum* may need to be separated.

Bibliography

ANDERSON, E. 1954. *Plants, Man, and Life*. Berkeley: University of California Press.

Anonymous. 1975. "The Gardeners Who Save Seed." *Organic Gardening and Farming* 22 (1): 54–55.

———. 1978a. "Saving Seeds." *Countryside* 62 (1): 46–48.

———. 1978b. "Tree Grafting Methods." *Organic Gardening and Farming* 25 (1): 82–96.

BAILEY, L. H. 1910. *The Principles of Vegetable Gardening*. New York: Macmillan.

———, and E. Z. BAILEY, 1976. *Hortus Third*. New York: Macmillan.

BERLIN, N. H. 1969. "Raising Strawberry Plants from Runners." *Organic Gardening and Farming* 16 (2): 51.

BOSWELL, V. R. 1949. "Our Vegetable Travelers." *National Geographic* 96 (2): 145–217.

BRIGGS, F. N., and P. F. KNOWLES. 1967. *Introduction to Plant Breeding*. New York: Reinhold.

BRINHART, B. 1969. "Florida Sunshine and Papaya." *Organic Gardening and Farming* 16 (12): 62–64.

CARL, R. A. 1975. "Seed Storage." *Mother Earth News* 31: 70–73.

CARLSEN, K. L. 1970. "Is It Time to Move the Rhubarb?" *Organic Gardening and Farming* 17 (4): 100–101.

COCHRAN, L. C., W. C. COOPER, and E. C. BLODGETT. 1961. "Seeds for Rootstocks of Fruit and Nut Trees." In *Seeds: The Yearbook of Agriculture*, U.S. Department of Agriculture, pp. 233–39. Washington, D.C.: Government Printing Office.

COX, J. 1974a. "New Vines from Grape Cuttings." *Organic Gardening and Farming* 21 (9): 44–47.

———. 1974b. "Top Grafting Step by Step." *Organic Gardening and Farming* 21 (10): 50–55.

DEAKIN, J. R., G. W. BOHN, and T. W. WHITAKER. 1971. "Interspecific Hybridization in *Cucumis*." *Economic Botany* 25 (2): 195–211.

DOUGLASS, B. 1976. "Go Full Circle in the Garden—Save Seeds." *Countryside* 60 (7): 43–44.

EMERY, CARLA. 1975. *Old Fashioned Recipe Book*. Kendrick, Idaho: Living Room Press.

FRANZ, M. 1973. "Climate Changes and the Origin of Plants." *Organic Gardening and Farming* 20 (12): 106–108.

GILLETTE, J. 1970. "Grafting Pecan Trees—Texas Style." *Organic Gardening and Farming* 17 (10): 46–49.

GOFF, E. S., and D. D. MAYNE. 1904. *First Principles of Agriculture*. New York: American Book.

HARTMANN, H. T., and KALE E. KESTER. 1975. *Plant Propagation: Principles and Practices*. 3rd ed. Englewood Cliffs, N.J.: Prentice-Hall.

HAWTHORN, L. R. 1961. "Growing Vegetable Seeds for Sale." In *Seeds: The Yearbook of Agriculture*, U.S. Department of Agriculture, pp. 208–215. Washington, D.C.: Government Printing Office.

HILLS, L. 1975. "Currant Affairs." *Practical Self-Sufficiency* 1 (1): 18–19.

JOHNSTON, R. L., Jr. 1976. "Seed Saving Is Fun: Here's How." *Organic Gardening and Farming* 23 (1): 60–63.

———. 1977. *Growing Garden Seeds*. Albion, Maine: Johnny's Selected Seed.

———. 1981. "Grow Your Own Garden Seeds." *Farmstead* 7 (8): 32–34.
KAHR, A. E., F. P. ESHBAUGH, and D. H. SCOTT. 1961. "Propagation of Crops without True Seed." In *Seeds: The Yearbook of Agriculture*, U.S. Department of Agriculture, pp. 134–44. Washington, D.C.: Government Printing Office.
KAINS, M. G. 1935. *Five Acres and Independence*. New York: Greenberg.
KING, R. 1977. "Save Your Own Garden-grown Vegetable Seed." *Mother Earth News* 47: 80–81.
KRUSE, O. 1969. "Easy Ways to Propagate Shrubs and Perennials." *Organic Gardening and Farming* 16 (7): 41.
LANGER, R. W. 1972. *Grow It!* New York: Saturday Review Press.
MARINER, R. 1977. "Horticultural Heirlooms." *Harrowsmith* 2 (2): 36–43.
McKAY, J. W. 1961. "How Seeds Are Formed." In *Seeds: The Yearbook of Agriculture*, U.S. Department of Agriculture, pp. 11–17. Washington, D.C.: Government Printing Office.
MEEKER, J. J. 1969. "Seeds from Your Own Garden." *Organic Gardening and Farming* 16 (7): 30–34.
MILLER, D. C. 1977. *Vegetable and Herb Seed Growing for the Gardener and Small Farmer*. Hershey, Mich.: Bullkill Creek.
MOFFETT, H. 1977. "The Black Walnut: Jupiter's Tree." *Country Journal* 4 (11): 46–48.
MOON, C. L. 1975. "How to Save Your Own Garden Seed." *Mother Earth News* 34: 14–15.
MOORE, F., and L. MOORE. 1975. "Home-grown Garden Seeds." *Mother Earth News* 34: 16–17.
NISSLEY, C. H. 1942. *The Pocket Book of Vegetable Gardening*. New York: Pocket Books.
RIOTTE, L. 1973a. "Jujubes: China's Gift to the Garden." *Organic Gardening and Farming* 20 (2): 122–23.
———. 1973b. "The Art of Growing Globe Artichokes." *Organic Gardening and Farming* 20 (3): 102–104.

———. 1974. "What's a Chayote?" *Organic Gardening and Farming* 21 (1): 158–59.

ROGERS, B. R. 1978. "Harvesting Your Own Garden Seeds." In *The Old Farmer's Almanac*, R. B. Thomas, no. 186, pp. 64–66.

ROZELL, B. 1970. "There's Nothing Wrong about Working for Peanuts." *Organic Gardening and Farming* 17 (4): 58–60.

SCHALES, F. D. 1969. "Never Plant Cucumbers next to ... " *Organic Gardening and Farming* 16 (5): 46–47.

SHADE, L. 1974. "Act Now to Save Money on Next Year's Seed." *Organic Gardening and Farming* 21 (9): 86–87.

STEELMAN, L. W. 1951. *A Few Acres and Security*. New York: Greenberg.

STOUT, R. 1971. "Do You Save Seeds? Watch It!" *Organic Gardening and Farming* 18 (1): 50–53.

TALBERT, T. J. 1946. *General Horticulture*. Philadelphia: Lea and Febiger.

THOMPSON, H. C., and W. C. KELLY. 1957. *Vegetable Crops*. New York: McGraw-Hill.

TIRRELL, R. 1969. "Perennial Onions–for Yields Every Year." *Organic Gardening and Farming* 16 (12): 33–35.

———. 1972. "Self-Sowing Vegetables Save You Work." *Organic Gardening and Farming* 19 (8): 53–55.

TRIPP, V. 1970. "Sweet Potatoes Are Prolific." *Organic Gardening and Farming* 17 (4): 86–87.

U.S. Department of Agriculture, Forest Service. 1974. *Seeds of Woody Plants in the United States*. Agricultural Handbook no. 450. Washington, D.C.: Government Printing Office.

VICK, E. C. 1928. "Good Vegetables and Market Gardening." *Audel's Gardener's and Grower's Guide*, Vol. 2. New York: Audel Educational.

WAHLFELDT, B. 1971. "Propagating Your Favorites by Sand-Rooting." *Organic Gardening and Farming* 18 (2): 54–55.

WILLMANN, O. 1977. "An Interview with Rob Johnston." *Farmstead* 4 (3): 44–47.

Index

almond, 40
annual plants, 8
annular budding. *See* ring budding
apple, 42
apricot, 42
artichoke. *See* globe artichoke; Jerusalem artichoke
asexual plant propagation, 13, 37. *See also* budding; grafting; plant division
asparagus, 45
avocado, 46

banana, 46–47
bark grafting, 34–35
barley, 47
bean, 47–49
beet. *See* red beet; sugar beet
biennial plants, 8

blackberry, 51
black cap, 119–20
black currant, 76
black raspberry, 119–20
black salsify, 123
black walnut, 134
blueberry, 52–53
bok choy. *See* pak choi cabbage
borecole, 90
bridge grafting, 31, 34
broccoli, 60, 61. *See also* sprouting broccoli
brussels sprouts, 54
buckwheat, 54–55
budding: advantages, 25; history of, 24; plants for, 27; methods, 27–29. *See also* budding and grafting
budding and grafting: choice of plants for, 25–26; general principles of, 25–

INDEX

budding and grafting (*continued*) 26; and growing rootstock, 26–27; tools and supplies for, 26. *See also* budding; grafting
bulblets, 23
bulbs, 21–22
butternut, 55

cabbage. *See* head cabbage; pak choi cabbage; pe-tsai cabbage
cantaloupe, 57–58
cardoon, 58
carrot, 58, 60
cauliflower, 60–61
celeriac, 61
celery, 62
chard. *See* Swiss chard
chayote, 63
cherry, 64
chervil, 64, 66
chestnut, 66
chicory, 67
chive, 67–68
cleft grafting, 30–31
coconut, 69
collards, 69–70
compound layering, 21
corn, 70
cowpea, 109
cranberry, 71–72
cress. *See* garden cress; watercress; winter cress
cross-pollination, 6
crosses, 3, 7
crown division, 16
cucumber, 73, 75. *See also* lemon cucumber
currant. *See* black currant; red currant
currant tomato, 131–32
cushaw pumpkin, 116
cuttings, 14–16

dandelion, 76, 78
dasheen, 131
date, 78–79

dewberry, 79–80
dioecious plants, 6
dormant stem cuttings. *See* hardwood stem cuttings

eggplant, 80
endive, 80
English walnut, 134–35

F-1 hybrids, propagation of, 7–8
field pumpkin, 116–17
fig, 81
filbert, 81–82
flowering plants: and male and female flowers, 6; seed production by, 4–6
flowers, parts of, 4–6
flute budding, 28
food plants, botanical classification of, 138–45. *See also* plant propagation; *and under specific names*
fruit trees. *See* budding and grafting

garden beet, 49–50
garden cress, 72
garden sorrel, 125–26
garlic, 82–83
genes, sexual propagation of plants and, 6–7
girasole, 44
globe artichoke, 43–44
golden thistle, 123–24
Good King Henry, 83
gooseberry, 83–84
grafting, 30–35: advantages of, 25; history of, 24. *See also* budding and grafting
grafting wax, 27
grape, 84
grapefruit, 86
green cuttings, 14
guava, 86–87
gumbo, 102

H-budding, 29
hardwood stem cuttings, 14–15
hazelnut, 81–82

INDEX

head cabbage, 56
hickory, 87-88
horseradish, 88-89
husk tomato, 132
hybrid "dropouts," 3, 8
hybrids, prevention of, 7

insect-pollinated plants, 6

Jerusalem artichoke, 44
jujube, 89-90

kale, 90
kohlrabi, 91-92
kumquat, 92

layering, 19-21
leaf mustard, 100-101
leek, 92-93
lemon, 94
lemon cucumber, 75-76
lentil, 94
lettuce, 94-95
lime, 96
loganberry, 79-80
loquat, 96

Macadamia nut, 96-97
mango, 97
manioc, 98
martynia, 98
millet, 99
monoecious plants, 6
mound layering, 20
muskmelon, 99-100

nectarine, 101
New Zealand spinach, 126-27

oats, 101-2
okra, 102
olive, 102-3
onion, 103-4. *See also* Welsh onion
orache, 105
orange, 105-6
ovules, 5
oyster plant, 123

pak choi cabbage, 56-57
papaya, 106-7
parsley, 107
parsnip, 108
patch budding. *See* flute budding
pea, 108
peach, 109
peanut, 110
pear, 110-11
pecan, 111
pepper, 111-12
perennial plants, 10
perfect flowering plants, 6
persimmon, 112-13
pe-tsai cabbage, 57
pineapple, 113-14
pistil, 5
plant division, 13-23
plant propagation: and F-1 hybrids, 7-8; history of, 1-3; selection of parent plants for, 36-37; types of, 4. *See also* budding; grafting; plant division; *and under specific names of food plants*
plate budding, 29
plum, 114
pollen, 5
pollinating agents, 6
pomegranate, 114-15
potato, 115-16
pumpkin. *See* cushaw pumpkin; field pumpkin

Queensland nut, 96-97
quince, 117-18

radish, 118-19
raspberry. *See* black raspberry; red raspberry
red beet, 49-50
red currant, 76
red raspberry, 120
rhubarb, 121
rice, 121-22
ring budding, 29
root "buds." *See* bulbs
root cuttings, 16

INDEX

root division, 17
root grafting. *See* whip grafting
rootstock, growing of, 26–27
runners. *See* stolons
rutabaga, 122
rye, 122

saddle grafting. *See* spliced grafting
salsify, 123. *See also* black salsify; Spanish oyster plant
sapota, 124
sea kale, 90–91
seed dormancy, stratification and, 12
seed production: commercialization of, 1–2; and flowering plants, 4–6; gathering, 10–11; and hybrids and crosses, 3; and identification of plants, 8, 10; for rootstock, 26–27; and selection of parent plants, 36, 37; and stratification, 12; and viability testing, 11
self-incompatible plants, 6
self-pollination, 6
self-sterile plants. *See* self-incompatible plants
sepals, 4
serpentine layering. *See* compound layering
sexual propagation. *See* seed production
shallot, 124–25
shield budding, 27–28
simple layering, 19
sorghum, 125
sorrel. *See* garden sorrel
southern curled mustard, 100
soybean, 126
Spanish oyster plant, 123–24
spinach, 127. *See also* New Zealand spinach
spliced grafting, 35
sprouting broccoli, 53–54
spur budding, 29
squash. *See* summer squash; winter squash

stamens, 4–5
stolons, 17, 19
stratification of seeds, 12, 26–27
strawberry, 129
suckers, 17
sugar beet, 50–51
summer squash, 128
sunflower, 129–30
sweet potato, 130–31
Swiss chard, 62–63

tapioca. *See* manioc
taro, 131
tip rooting, 21
tomato, 132–33. *See also* currant tomato; husk tomato
top grafting. *See* cleft grafting
trees, hardwood stem cuttings, 14–15
true-breeding plants, 6–7
"true" seed, 37
tubers, 22–23
turnip, 133–34

unicorn plant, 98

vegetable oyster, 123
vegetable pear. *See* chayote
veneer budding. *See* flute budding
viability testing of seeds, 11

walnut. *See* black walnut; English walnut
watercress, 72–73
watermelon, 135
Welsh onion, 104
wheat, 135–36
whip grafting, 30
whip and tongue grafting. *See* whip grafting
wind-pollinated plants, 6
winter cress, 73
winter crookneck, 116
winter squash, 128–29

yam, 136

NEW HANOVER COUNTY
PUBLIC LIBRARY

631.52
F
　　Fitz, Franklin Herm
　　A gardener's guide to
　　propagating food plants

NEW HANOVER COUNTY PUBLIC LIBRARY
WILMINGTON, NORTH CAROLINA